Wastewater Engineering

Third Edition

J.B. White

MSc Tech, FICE, FIWEM
Senior Lecturer in Civil Engineering
University of Manchester Institute of Science and Technology

Edward Arnold

First published in Great Britain 1970 as the Design of Sewers and Sewage Treatment Works by Edward Arnold (Publishers) Ltd, 41 Bedford Square, London WC1B 3DQ

Edward Arnold (Australia) Pty Ltd, 80 Waverley Road, Caulfield East, Victoria 3145, Australia

Edward Arnold, 3 East Read Street, Baltimore, Maryland 21202, U.S.A.

Reprinted 1974
Second edition 1978
This edition published 1987

British Library Cataloguing in Publication Data

White, J.B.
 Wastewater engineering.——3rd ed.
 1. Sewerage——Great Britain 2. Sewage
 disposal plants——Great Britain
 I. Title
 628'.2'0941   TD557

 ISBN 0-7131-3614-6

Text set in 10/11pt Baskerville Compugraphic
by Colset Private Limited, Singapore
Printed and bound in Great Britain by
The Bath Press, Avon

Preface

The objective of this book, as in previous editions, is to give an introductory and explanatory account of the design of sewers and sewage treatment works. It is intended for engineers beginning to practise in this field and contains more practical detail than it is possible to include in a course of lectures. However, I have aimed at including full explanations of basic principles since a proper understanding of these is an essential foundation.

Much literature is now available due to the expansion of officially funded research and development. The present text is no substitute for such material but should serve to place it in perspective and to offer the basic explanations which would not be appropriate in manuals written for engineers already in established practice.

All chapters have been revised, though much has been retained from the two earlier editions. Much of Chapter 2 has been re-written to describe the Wallingford Procedure methods for assessing urban run-off, and the opportunity has been taken to include an account of the use of balancing ponds. A fuller account of storage overflows has been given in Chapter 4 and material on sedimentation tanks — both primary and final — has been modified in Chapter 10.

Chapters 6 and 7 divide the chapters on sewers from those on sewage treatment and are applicable to both aspects of wastewater engineering. These chapters have been revised and extended for the present edition and calculations exemplifying the material have been added to Chapter 14 on the overall design of treatment works.

Illustrations of proprietary equipment are included and I am indebted to the manufacturers, acknowledged below each illustration, who have allowed me to use their material. Most of these firms, and some others, manufacture a full range of equipment and should be consulted at an early stage of design to obtain the latest information on their products.

Though some of the bibliographical references given at the ends of chapters are to works which are in constant use in design offices, others are given in order to acknowledge the source of a particular item of information. Many of these papers contain much more of interest and are well worth reading in entirety, especially when an answer is being sought to a particular problem.

The first edition would not have been written without the encouragement of the late Professor J.R.D. Francis and I remain grateful for the help he gave. Much is also owed to a great many students with whom I have exchanged ideas and tried out explanations. The many engineers who have attended the short courses I have given in recent years have enabled me to keep in touch with developments and have thereby taught me much of value.

It is too much to hope that I have managed to eliminate all erors from the text. I am grateful to readers who have drawn my attention to errors in the earlier editions.

JBW
Manchester
July 1986

Contents

1

Introduction

1.1 The origins of present-day practice

The need for systems of water sanitation in large towns became evident
during the Industrial Revolution. Urban areas were developed without
adequate provision for water supply or for the removal of waste. Water was
taken from shallow wells, from polluted streams or, at best, from leaky water
mains which were kept under pressure for only a few hours each day.
Accumulations of waste matter resulted in the contamination of water
supplies. High mortality from the water-borne diseases, typhoid, cholera
and forms of dysentery, was widespread in the densely populated areas.

As Secretary to the Poor Law Commission, Edwin Chadwick[1] realised
that much poverty was the result of disease and early death. Though it was
not then known that bacteria were responsible for the spread of diseases such
as cholera, Chadwick became convinced that the failure to remove waste
matter promptly and the lack of clean water supplies had some connection
with the prevalence of disease. Chadwick and his supporters campaigned
energetically for improvement. The report of 1841 on the Sanitation of
Towns contains descriptions of conditions which can scarcely be believed
today.

The solution proposed by Chadwick is the one now familiar. Each
dwelling is supplied internally with clean water, continuously under
pressure, and the water used conveys waste matter through a system of pipes
at velocities high enough to prevent silting. Previously, sewers had been
constructed of stone and brick. Silting and blockage were common and
sewers were made large enough to accommodate deposits. John Roe had
shown that deposition could be avoided by using 'tubular' or pipe sewers of
smaller cross-section provided that the gradients were steep enough.

The proposals put forward by Chadwick were embodied in schemes
produced as a result of the Public Health Act of 1848. This act was repealed
later but subsequent legislation followed similar lines.

The only method of sewage treatment envisaged at this time was the distribution of sewage over an area of land at a safe distance from the town. It
was considered that this would improve the land agriculturally. Very large
areas of land were needed to prevent conditions from becoming obnoxious

and attempts were gradually made to find better methods of treatment. Later in the century a Royal Commission was appointed to review the problems and the available information. The recommendations in the reports of the Royal Commission ultimately provided what amounted to a code of rules for sewage works design. These greatly influenced British practice and, while many departures have been made since, vestiges of the Royal Commission's recommendations remain to the present day.

1.2 The public authorities

Before 1974 the provision of sewers and sewage treatment works in England and Wales was the responsibility of the Councils of Municipal and County Boroughs and of Urban and Rural District Councils. Capital for construction was obtained in the form of loans from central government repaid with interest by equal annual instalments as a charge on the rates levied on property values by the authorities. There was a measure of central control of technical standards in that the appropriate Ministry — latterly the Department of the Environment — considered schemes for purposes of loan sanction. In addition the Local Authorities had to gain the approval of the River Authority for discharges made from treatment works and storm overflows.

The Water Act of 1973 which came into force on 1st April 1974 brought about a radical change transferring sewerage and sewage treatment responsibilities to ten new regional Water Authorities. Simultaneously the 1973 Local Government Act set up a two-tier system of County and District Councils. The new District Councils and some of the new County Councils covered areas larger than those of their predecessors. Though the primary responsibility for sewers now lies with the Water Authorities provision was made for District Councils to act as their agents. Currently, with few exceptions, the sewerage work is carried out by District Council staff but it is subject to control by the Water Authority and expenditure forms part of the Water Authority's overall programme. Sewage treatment is not delegated to Local Authorities.

The Regional Water Authorities were set up so that all aspects of water use could be planned and controlled over much wider areas of the country than before. This appeared to be necessary because, to meet increasing demands for water, lowland rivers would have to play a more significant role in water supply. In addition to making more use of the natural flow of such rivers, their resources would be augmented by upland regulating reservoirs, by river-to-river transfers and by schemes for the conjunctive use of ground water storage. When schemes of these kinds are envisaged the quality of the river water becomes a crucial factor in water supply and it was considered that Water Authorities should be created with powers which included the provision and operation of sewage treatment works.

In Scotland water services (water supply, sewerage and sewage disposal) are primarily the responsibility of the upper tier of a mainly two-tier system of general purpose Local Authorities created on 16th May 1975. The upper tier consists of nine Regional Councils and three Islands Councils. However, rivers pollution prevention is the responsibility of seven River Purification Boards except in the islands where the three Islands Councils act also as River Purification Authorities.

In Northern Ireland, sewerage and sewage treatment (and other water services) have been administered centrally by the Ministry of Development since April 1973.

1.3 Combined and separate sewers

Sewerage systems are designed to drain surface water as well as polluted water from built-up areas. Originally a single system of pipes known as a 'combined system' was used to convey both surface and polluted water. Later, in new developments, two sets of pipes were used, surface-water sewers for the rainfall drained from paved and roofed areas, and foul sewers for polluted water. This is known as the 'separate system' of sewerage. In order to save some of the extra expense of a completely separate system, a compromise known as the 'partially separate system' was adopted in some areas. On this system, surface-water sewers deal with the greater part of the road and roof drainage while a second set of pipes takes all the polluted water together with surface water from yards and rear roofs which are close to the sources of polluted water.

The 'combined system' was not deliberately adopted but developed in a haphazard fashion. As areas were built upon, natural watercourses were culverted and house drains were connected to them. The culverts discharged into the nearest river. As large towns grew up, river pollution became intolerable and intercepting sewers were constructed parallel to the river to convey sewage to a remote area downstream where land treatment could be given to the sewage.

Even in the United Kingdom where the maximum intensities of rainfall are relatively moderate, the maximum rate of surface-water flow from a sewered area is 40 to 150 times the average rate of flow of polluted water, the factor being larger for smaller areas. It was evidently uneconomic to make intercepting sewers large enough to take all the flow from existing sewers, so storm overflows were constructed at the points of connection with the intercepting sewers. Flows in excess of the capacity of the intercepting sewer were diverted to the river. This was justified on the grounds that the sewage would be greatly diluted during rainstorms.

Where a town was already sewered on the combined system, new areas of development were provided with deliberately designed combined sewers, storm overflows being installed wherever a watercourse ran near to the system so that pipe diameters could be minimised. It became customary, following Royal Commission recommendations, to design the overflow chambers so that through-flow to the treatment works was limited to six times the estimated mean rate of foul flow in dry weather, expressed briefly as 6 × DWF.

The use of storm overflows often caused stream pollution, especially when a heavy storm washed out deposits which had accumulated during long dry periods. This led to the adoption of the separate system for further areas of housing development. The separate system is more costly as the surface-water sewers are themselves as large as combined sewers and a foul system has to be provided in addition. Two sets of house drains have to be provided and external plumbing is more elaborate as bathroom wastes cannot be discharged into rainwater pipes.

The surface-water pipes of a separate system usually discharge to a nearby watercourse at a point where a storm overflow outlet would have been provided on a combined system.

The two service sewers of a separate system are laid in a common trench, the foul sewer being alongside and at a lower level than the surface-water sewer so that side connections can be accommodated. A common trench reduces total excavation costs and enables twin-chamber manholes to be used.

The separate system has been used for many areas of housing development constructed since about 1920, though in some places the system is only partially separate. Where rural areas are provided with sewers, the system is usually separate since pipes are laid for foul sewage alone, existing arrangements being left to deal with surface water. The opportunity may be taken to connect farmyard drainage to the sewers.

Where rainfall intensities are high as in tropical and monsoon climates, the combined system is scarcely feasible and sewers are provided for polluted water only. Much of the surface water is conveyed by open ditches which are not regarded as part of the sewerage system.

Combined systems usually occur in the more densely developed parts of towns where the cost of conversion to the separate system would be prohibitive. However, in areas of redevelopment, substantial lengths of new service sewers are often needed and existing sewers are abandoned. Reconstruction of trunk sewers has to be undertaken from time to time. A long-term programme of gradual conversion may therefore be feasible if each redevelopment area is sewered on the separate system.

The combined system is not without its advocates and it is undoubtedly true that, even now horse-drawn traffic has disappeared, there are paved areas which ought to be drained to a treatment works. There can be little harm in using the separate system for residential areas. It is true that grit and other deposits washed from roads would reach the river and might aggravate silting, but road gullies trap a good deal of grit, and catch-pits can be installed where the serious problems are likely to arise.

In commercial and industrial areas, spillage from tank vehicles or processing plants and accumulation of waste material present a hazard and many paved areas other than roads should be drained to the foul sewers on what amounts to a partially separate system. Whether the roads themselves should be drained to a treatment works depends on individual circumstances. In some heavy industrial areas a good case can be made for the use of the combined system, especially since the design of storm overflows has been considerably improved.

1.4 General features of sewerage systems

Most sewers are laid under roads to avoid interference with private property when connections and repairs have to be made. These reasons do not apply on municipal housing estates but even there the road is often the most convenient place for a surface-water or combined sewer to which road gullies are connected. Figure 1.1 gives a typical layout for sewers in residential areas.

Sewers are commonly laid in straight lines, manholes being provided at all changes of direction, gradient and diameter. This was a specific requirement for loan sanction approval as also was a requirement to provide manholes at distances of not more than 110 m on long straight lengths (this distance may be increased for sewers which are large enough for men to work in). These requirements are primarily to facilitate rodding when blockages occur. In addition they enable the sewer line to be precisely located whenever the sewer has to be exposed for connections or repairs.

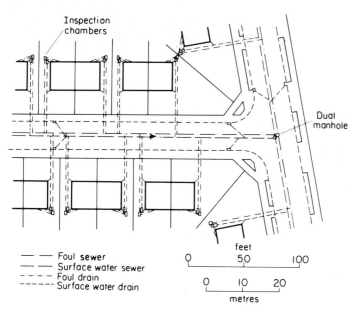

Fig. 1.1 Housing area drained on the separate system. Some expense could be saved by using common drains for adjacent houses. This might be done on municipal housing estates

Some engineers believe that 110 m is too large a minimum for small-diameter pipes and provide manholes at say 80 m intervals on straight lengths. An old rule for large sewers is that the minimum manhole spacing should be 100 yd for each foot (300 m per m) of diameter. However, it is probably better to provide manholes at intervals of not more than about 300 m even on very large sewers. It is worthwhile to encourage regular inspection by making access easy and by ensuring that manholes are sufficiently close to provide adequate ventilation. Manholes account for a very small proportion of the cost of large sewers, unless they are very deep, and collapses of disastrous proportions have occurred in sewers which were virtually impossible to inspect because of the lack of adequate access.

It is arguable that the practice of laying sewers in straight lines has been followed too rigidly in the United Kingdom and experience elsewhere has shown that sewers laid on gentle curves are not unsatisfactory. Sewers large enough for man access have sometimes been curved on plan and a Ministry Working Party[2] recommended that this should be considered acceptable. Laying sewers on curves saves manholes, though distances should still be limited so that rodding and inspection are easy. Careful records of the sewer line *as laid* must be kept if the line is not straight.

Most service sewers are laid at depths of 1.5 to 2 m. This is normally sufficient to allow adequate gradients in the connecting drains though where there are basements or where premises are further from the road or at a lower level, the sewer has to be deeper. At depths of cover less than 1.5 m under roads, the effect of traffic loads is increased and the provision of extra strength in construction can outweigh the saving in depth of excavation.

References

1 Finer, S.E., *The Life and Times of Sir Edwin Chadwick*, Methuen, 1952.
2 Ministry of Housing and Local Government, *Working Party on the design and construction of underground pipe sewers, 2nd Report*, H.M.S.O., 1967.

2

Design of sewerage systems

2.1 Preliminary steps

The first stages in designing a system of sewers are the same for combined, separate and partially separate systems.

On a large-scale map of the district, a suitable scheme of pipe-runs is drawn up. If contours are not available, the site has to be visited first and the directions of slope of all roads and land adjacent to roads marked on the plan. Details of existing sewers and drains are collected.

The principles governing the layout of sewers have been outlined in Chapter 1. The sewers will follow the roads in most cases and will form a system which branches in the upstream direction from the point of outfall. Often, several possibilities will present themselves but, with a little experience, the best one can be chosen without making flow calculations for detailed comparison.

After the layout has been settled, ground levels will be taken along all sewer lines so that precise gradients can be worked out. The ground levels will be used later for contract drawings. Occasionally it is found necessary to modify the original layout when the actual levels are considered.

The next stage is to consider each individual length of sewer (between manholes) and to define the area which can be drained to it. These are the smallest units of area used in flow calculations: it is not considered necessary to subdivide them into the areas draining to individual gully or drain connections. On the separate and partially separate systems it is advisable to prepare two drainage area plans, as the areas contributing to the surface-water sewers do not always coincide with those contributing to the sewers taking the foul sewage.

When calculations are to be done to check an existing system, the areas have to be defined according to the actual drain connections and some tests in the field using dyes for tracing purposes may be needed to identify the sewers to which particular properties drain.

A standard system of numerical references for sewer lengths is adopted in design calculations. The system was devised for use in computer programming to define the branching relationships of successive pipes, but it is commonly used whether or not the calculations are to be done by computer.

Each length of sewer between adjacent manholes is given a reference of the form *a.b* in which *a* is a number relating to the particular branch of the layout and *b* is a number defining the length along the branch. Numbering is begun at the top end of the longest branch with the reference 1.1. The next length downstream is called 1.2 and so on until a junction is reached. The branch entering at the junction is numbered 2 and its pipes working downwards are 2.1, 2.2, 2.3, etc. Proceeding onwards from the junction, the first branch number, 1, is used for the main line. An example is given in Fig. 2.1. Pipe numbering is sometimes begun at 0 instead of 1.

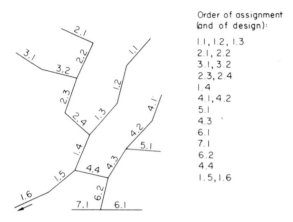

Order of assignment
(and of design):

1.1, 1.2, 1.3
2.1, 2.2
3.1, 3.2
2.3, 2.4
1.4
4.1, 4.2
5.1
4.3
6.1
7.1
6.2
4.4
1.5, 1.6

Fig. 2.1   Referencing system for sewer lengths. The references listed in the order of assignment define the form of the layout

In design, whatever the method, the pipes are dealt with in the order in which the numbers have been assigned. Whenever a branch is encountered, its pipes are designed before work on the main line is continued. The references, listed in the order of assignment, define the shape of the system and data for use in computations are listed in this order.

For contract purposes it is usual to number the manholes rather than the pipes and numbering is started at the lower end so that the manholes are numbered roughly in the order of construction. Working uphill simplifies drainage during construction. It is not worthwhile to attempt to use the same system for both calculations and contract drawings as manholes are sometimes added or eliminated at a late stage.

2.2 Estimating flows and choosing pipe sizes

For each pipe the maximum likely flow has to be estimated and a diameter chosen to suit the gradient available. Where pipes receive connections between manholes it is the flow at the downstream end which is important.

Diameters are chosen so that the estimated flow will be carried with the pipe running 'just-full', that is the pipe is assumed to be full but not sur-charged and the hydraulic gradient is the same as the invert gradient. This makes the calculations simple because part-bore flow does not have to be considered. Due to the properties of circular sections (Fig. 6.4), a pipe will actually flow at about 80% of full depth with the design discharge and there will be a margin of about 7% available above the design discharge before the pipe becomes surcharged. Sewers can often flow in a surcharged condition without causing flooding so there is sometimes an even greater safety margin.

Pipe gradients must not be so flat as to give running full velocities of less than 0.9 m/s (or 0.75 m/s for pipes of 0.75 m or more in diameter). See Section 6.8 for further discussion of velocity limits.

2.3 Flows of foul sewage

Almost all water supplied from the public mains for domestic purposes is dis-charged to the sewerage system. In a residential area, therefore, water supply records are the starting point for estimates of flows in sewers. Figures for domestic water consumption on a per head per day basis are usually readily available, but the source of these figures needs some explanation as with rare exceptions water supplied to houses is not metered individually.

Most water supply systems are divided into districts, meters being installed on the mains supplying each district, but many districts contain trade premises with supplies metered individually. Consequently it is possible from water records to estimate the supply to domestic consumers by subtracting the total of the individually metered supplies from the quantity supplied to the whole district. This will commonly be an overestimate of the water actually supplied to consumers since the loss of water by leakage from the distribution mains is significant. In many areas waste detection and pre-vention are actively pursued and the water supply engineer will often have a fairly accurate notion of the amount of waste.

The figures most frequently quoted for domestic water consumption per head per day usually include the wastage, and are often used as average sewage flow per head per day with the recognition that they are a safe over-estimate. Typical values range from 140 to 230 litre per head per day.

For buildings occupied only during the day such as schools, offices and communal premises, and for hospitals and residential institutions, water supply records are again the most useful means of assessing average sewage flow. Data on a per head per day basis are needed for estimating flows from proposed developments.

Water supply records may not be adequate by themselves for manu-facturing premises. Additional water may be being used from private sources such as boreholes. Not all the water may be discharged to the sewers; some may be used up in evaporation or become part of the product and some may be discharged to a river. Where the factory is making licensed discharges records should be available, otherwise each case must be investigated.

Sewers have to convey infiltration water leaking into them from the ground. In old sewers infiltration can be of the same order as domestic sewage flow. It is estimated by gauging flows during the early hours of the morning.

Water supply records, data from factories and infiltration estimates enable the daily total dry-weather flow to be obtained and this value, in effect the mean rate of flow, is usually referred to as the DWF (when the initial capital letters are used the mean value is implied). During the day, flow fluctuates about this mean value reaching a maximum of 2 to $2\frac{1}{2}$ times DWF during the middle part of the day and having a minimum of $\frac{1}{4}$ to $\frac{1}{2}$ of DWF during the night consisting mainly of infiltration. There is often some seasonal variation and a characteristic weekly pattern in addition to the diurnal variation. For sewers serving small groups of premises the maximum will be higher than $2\frac{1}{2}$ times DWF as there will be less balancing of discharges.

2.4 Design flows for foul sewers (separate system)

The Code of Practice for Sewerage provides design guidance but needs inter-preting with some care.[1] To begin with the code states that the foul sewers of a rigidly separate system 'shall be capable of carrying not less than four times the average dry-weather flow; and allowance of six times is not excessive, but this figure is seldom exceeded by designers'. Since it is clear from the definition section that dry-weather flow includes infiltration water the implication is that both the mean sewage flow and the mean infiltration flow are to be multiplied by four (or six). This appears to be an over-generous allowance for infiltration. However, it is later stated that in the absence of clear and reliable evidence the average domestic dry-weather flow should be assumed to be not less than 136 litre and generally not more than 227 litre per head per day depending on the type of property. These are the figures which were given in the first edition of the code where it was further stated that with properly constructed sewers no allowance need be made for infiltration, except where existing sewers with infiltration flow were connected, or where ground conditions, such as mining subsidence, were unusual. The 1968 edition of the code is less confident of the absence of infiltration and recom-mends that an estimate of infiltration should be made and allowed for where substantial lengths of sewer will be below the water-table.

Since the figures given in the code for average domestic dry-weather flow are closely similar to those for domestic water supply the practical implica-tion seems to be that one multiplies these by four (or six) and then adds an allowance for infiltration if considered necessary in local circumstances. This would be consistent with the Report of the Technical Committee on Storm Overflows (see pp. 86 and 311) where multipliers of DWF are applied only to the mean sewage flow, mean infiltration being added separately.

2.5 Combined and partially separate sewers

Since both these types of sewer convey surface water, design procedure is similar to that for surface water sewers, an additional allowance being made for the foul flow. The latter is a very small fraction of the total for combined sewers and is sometimes neglected. The maximum through-flow from storm overflows is quite high in comparison with the surface water entering downstream, and should always be taken into account. The flow of foul sewage is significant also in partially separate sewers where the sources of surface water are relatively small.

2.6 Surface-water flow

The problem of estimating flows of surface water in order to design sewers has something in common with the problem of estimating flood flows in rivers. The river engineer has an intrinsically more difficult problem to solve since the features of a river and its catchment defy numerical representation except in very broad terms, whereas sewers are of regular form and slope and they drain areas which are both easily defined and highly impermeable. However, the river engineer has some advantages; he can estimate floods simply by studying records of flow of the river in question or of comparable rivers. Only in the absence of long records does he have to fall back on attempting to estimate river flow from rainfall. Even then comparatively short records enable the rainfall to run-off transformation to be deduced empirically, whereas the sewerage engineer has to find some tractable and thereby unrealistic way of representing this process in a theoretical manner.

The earliest methods of calculating sizes for surface water sewers evolved into the so-called 'Rational' method. Criticisms of this method arise from the extreme simplicity of its assumptions regarding the transformation of rainfall to run-off. However, the method makes plain that the choice of storm duration, because of its inverse relationship with maximum rainfall intensity, is of prime importance in assessing maximum flows.

The ability of computers to do any conceivable amount of arithmetic has meant that very complex models of run-off generation can be used to compute complete hydrographs and that, incidentally, a series of storm durations can be tried out to find the worst effect.

Recent work on hydrograph methods[2] has shown that the Rational method, with some improvements in accuracy and flexibility, is still useful for designing surface water sewers up to 1 m in diameter, so it remains the routine mode of design for a great deal of new work.

2.7 The Rational method

During a rainstorm the intensity, measured as depth of rain per unit time, varies with time and over the storm area. Frequently the intensity reaches a

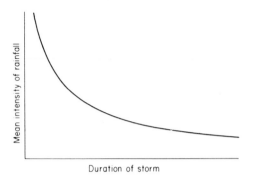

Fig. 2.2   Mean intensity of rainfall versus duration of storm

maximum near the middle of the storm period and near the middle of the storm area. Many storms are not stationary but move across country. For the purpose of the rational principle all these variations are ignored. In using the results for flow calculations it is assumed that the peak flow produced by a storm of given uniform intensity and duration will differ negligibly from that produced by a varying storm with the same rainfall depth and duration.

Mean intensity for a storm is the depth of water accumulated, divided by the duration of the rainfall. After data have been collected from recording rain gauges over a number of years they can be plotted on a graph relating depth of rain to duration of fall or, for the present purposes, relating mean intensity to duration. Each individual storm gives a point on the graph, see Fig. 2.2. The points can be contained within a curve as shown, though for the curve to be regular a few points may have to be left above it. Indeed a family of curves may be drawn such that each has a selected number of points, n, outside it. Each curve then represents storms equalled or exceeded at mean intervals of time of N/n years (the return period), N being the length of the record period in years. For practical use mathematical expressions describing such curves are obtained by statistical analysis.

The noteworthy property of intensity–duration curves is that 'small showers last long but sudden storms are short'.[3]

Next the manner in which flow increases during a uniform storm must be considered. This is illustrated in Fig. 2.3. The steady rate of rainfall input is Ai in which A is the impermeable area drained and i is the uniform rainfall intensity. At the beginning of rainfall the flow in the sewer is zero. At time t_S, flow, Q, at the outfall has become equal to the input rate Ai and the system has reached a steady state. Steady flows are then passing over all the drained surfaces and through all the pipes. The volume of water on the surfaces and in the pipes is equal to the area abc on the graph. This water drains away at the end of the storm, flow declining along the 'recession curve' de. Not quite all the water drains away since the surfaces are left wet when flow over them ceases, or the area def will be a little less than the area abc unless the surfaces

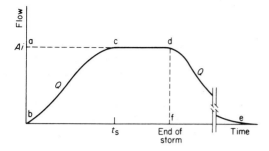

Fig. 2.3 Sewer-flow hydrograph (bcde) caused by hypothetical uniform rainfall input (badf)

are in the same state of wetness at the beginning and at the end of the flow. The curve bc is termed an '*S*-curve'. The time t_S may well vary with i but for the next step in the argument it is assumed to be constant.

On Fig. 2.4 flow hydrographs for a given case with fixed t_S are shown for uniform storms of intensity i_1, i_2 and i_3. The duration of each storm is obtained from a graph like Fig. 2.2 for a chosen return period. Storm (1) having a low intensity has a long duration t_1 and there is ample time for outflow to reach input rate Ai, and to remain at that value until rainfall ceases. It is evident that higher intensities would produce higher flows though the duration of steady flow would progressively shorten until at an intensity of i_2 for which $t_2 = t_S$ the steady state will occur only for instant as rainfall ceases as the flow Ai_2 is reached.

In storm (3), though the intensity is very high, the storm is so short that

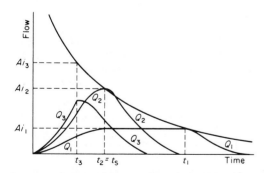

Fig. 2.4 Rainfall input and sewer-flow hydrographs for uniform storms of different duration and intensity. The intensities of rainfall i_1, i_2, and i_3 correspond to storms of duration t_1, t_2 and t_3, all of the same frequency and result in the outflow hydrographs Q_1, Q_2, and Q_3. The highest peak flow is given by the storm whose duration t_2 is equal to the *S*-curve rise time, t_S. Note that if the early part of the *S*-curve had risen more swiftly a storm of shorter duration than t_S might have produced a higher peak (cf: tangent method)

outflow, Q_3, is far below Ai_3 when rainfall ceases and flow begins to decline. Indeed the maximum flow reached is less than that in storm number 2.

Whether all storms (of the given frequency) with intensity greater and duration less than storm (2) give a lower maximum depends on the shape of the S-curve. In applying the rational principle it is assumed that the S-curve is not of such a shape that storms shorter than t_S give higher maximum flows.

The rational principle may be stated: for storms of given frequency the highest flow is given by the storm whose duration is equal to the S-curve time for the catchment. It follows that the storm causing the maximum flow in a river or sewer differs with the size of the catchment. For sewers and small streams short sharp thunderstorms are critical while long, steady, less intense rains give flows less than the maximum. On the other hand short thunderstorms may have only a slight effect on the flow of a large river while several days of lighter steady rain will set up a flood. Even for points not very far apart along a river or sewer different storms may be critical and indeed when the rational principle is applied in design a different storm is used as the critical case for each separate point at which an estimated peak flow is required.

Though the rational principle deals only with hypothetically uniform storms its implication that there is a relationship between the duration of the critical severe storm and the size of catchment must still apply with real storms of varying intensity.

When the rational principle was evolved the time taken from the beginning of uniform rainfall for flow to reach the steady state, t_S, was considered to be virtually the same as the time of travel of flow from the furthest boundary of the catchment area. This period was termed the 'time of concentration'. In practice the time of travel is taken as that when the pipes are flowing nominally full. Time of concentration is thus expressed as

$$t_c = t_f + t_e \qquad\qquad (2.1)$$

in which t_f is the summation of times of flow along the route of sewers at running-full velocities, and t_e, the time of entry, is an allowance for the time taken for flow to reach the head of the sewer from the furthest point drained.

Recommendations[2] for t_e relate it to return period, T, as follows:

t_e	3 to 6	4 to 7	4 to 8	5 to 10	minutes
T	5	2	1	$1/_{12}$	years

The shorter times are for areas less than 200 m² and steeper than 1 in 30; the longer times are for areas greater than 400 m² and flatter than 1 in 50.

The procedure, using the rational principle, is to take the time of concentration at each point where a design flow is required as the duration of the critical storm in order to obtain a rainfall intensity to be applied to the area draining to that point. Where a pipe receives flow from branches along its length, a design flow is needed at its lower end, so t_c includes the time of flow along the pipe to be designed and a trial of pipe diameter has to be made.

It will be appreciated that the time of concentration concept is an over-simplification of the *S*-curve process. In reality the sewers slowly fill, the inflow of storm water being more than the steadily increasing outflow. Neglect of a calculated allowance for the water temporarily stored in the pipes and on surfaces has been regarded as a major shortcoming of the Rational method.

2.8 Return period

The return period, in the context of sewer design, is the average length of time between occasions when a storm flow becomes equal to or exceeds the capacity of the pipe. The longer the return period used in the design and the larger the capacities and hence the cost of the pipes.

Compared with those used in other branches of hydraulic design, return periods adopted for surface water sewers are very short, a great many designs in the past having been based on rainfall return periods of no more than one year. The consequences of design flows being exceeded are often small, local and fairly brief. Many sewers can carry extra flow when surcharged without visible sign on the surface. Road gutters can carry significant additional flow at the height of a storm.

The return period of a given rate of flow in a surface water sewer is not necessarily the same as that of the rainfall used to calculate it. Because of factors such as the state of wetness of the ground when the rain starts, a range of storms can produce similar peak flows and conversely similar storms can produce dissimilar peaks of flow. There are grounds for believing that flows calculated by the Rational method have longer return periods than those of the rainfall intensities used.

The return period to be used for a design could be determined objectively by comparing the extra costs of providing larger pipes with the savings arising from the expected reduction in flood damage but there are many uncertainties so judgment cannot be entirely eliminated. The tendency is to use longer return periods for flat than for steeper districts and still longer periods for commercial districts where the flooding of basements would cause expensive damage. Long return periods should always be considered for critical points in the system such as pumping stations and inverted siphons.

The return period chosen has a very large effect on the flow calculated. Compared with a one-year return period, a 10-year period roughly doubles the flow, and a 100-year period roughly trebles it. An approximate rule for the factor is thus

$$1 + \log_{10}(\text{return period})$$

2.9 Rainfall intensity

The inverse relationship between mean rainfall intensity, I, and duration of storm, D, can be expressed simply as

$$I = a/(b + D) \tag{2.2}$$

in which a and b are constants obtained by fitting the expression to rainfall data. For many years the Ministry of Health formulae were used in Britain for sewer design. In these, pairs of values of a and b were given for different ranges of D. Later, further analysis of records showed that the formulae related to a return period of about one year and the more complex Bilham formula, offering a choice of return period, was adopted. Versions of both types of formulae have been developed for use in various parts of the world.

For the British Isles, in the course of work for the N.E.R.C. Flood Studies Report,[4] use was made of a formula as follows

$$I = I_0/(1 + BD)^n \tag{2.3}$$

in which I_0, the limiting value of I when D becomes zero, B and n could be assigned different values according to geographical location. These were related to r, which, in the notation of the Report is the ratio: $(M5\text{-}60 \text{ min})/(M5\text{-}2 \text{ day})$ where $MT\text{-}D$ signifies the maximum depth of rainfall in a storm of duration D and return period T. Values of r had a fairly close correlation with annual average rainfall.

A further step was taken in evolving the Wallingford procedure by developing a formula which used r directly, together with $M5\text{-}60$ min. In the Modified Rational method version, a factor, z_1, enables rainfall depth to be obtained as

$$(M5\text{-}D) = (M5\text{-}60 \text{ min}) \times z_1$$

A graph gives z_1 for values of r and for durations up to 48 hours. A map shows how $M5\text{-}60$ min varies across the country.

If calculation is preferred to reading z_1 from the graph, the following expression for z_1/D can be deduced

$$\ln(z_1/D) = \left(\frac{\ln(721/(1 + 15D))}{\ln(721/16)} - 1 \right) \ln \frac{48r}{1.06} \tag{2.4}$$

Intensity for the flow calculation is then obtained from:

$$I = (M5\text{-}60 \text{ min}) \times (z_1/D) \tag{2.5}$$

Equation (2.4) is based on (2.3) but uses $M5\text{-}60$ min and r as parameters instead of I_0 and n with $B = 15 \text{ hr}^{-1}$.

To convert $M5$ values into those for another return period, a factor z_2 is used, formulae and tables for which are given in the Wallingford procedure. Though z_2 varies with rainfall depth, the variation is very small for the return periods commonly chosen for sewer design and does not lead to any additional complication.

Data from which rainfall intensity formulae are developed are from rain gauges giving point values of rainfall depth. In places with a suitably close network of gauges it has been possible to work out average or 'areal' depths over various portions of the area and for various durations of storm. When this has been done, it has been found that areal depths of a given return period are always less than point depths for the same return period. The ratio of areal to point depth is termed the areal reduction factor or ARF. It decreases with extent of area and with duration of storm.

Strictly, for run-off calculations, the areal rainfall rather than point value should be used, so intensities from the formulae above should be multiplied by ARF. For river catchments, ARF is quite significantly less than one but for the small areas and short durations met with in most sewerage schemes it is very close to one and varies so slightly that a common value can often be used for several successive pipes.

The Wallingford programs use the following to calculate ARF, for *total* area A in km^2 and duration D in hours:

$$ARF = 1 - f_1 D^{-f_2} \tag{2.6}$$
$$f_1 = 0.0394\, A^{0.354} \qquad\qquad A < 100$$
$$f_2 = 0.40 - 0.0208 \ln(4.6 - \ln A) \qquad A < 20$$
$$f_2 = 0.40 - 0.00\,382\,(4.6 - \ln A)^2 \qquad 20 \leqslant A \leqslant 100$$

2.10 The Modified Rational method

In this method the design flow for each pipe is calculated from

$$Q = C_v C_R I_a A / 360 \tag{2.7}$$

in which Q is in m^3/s, I_a in mm/hr, A in hectare (ha) and the coefficients C_v and C_R are dimensionless.

The areal rainfall intensity, I_a, is found for the chosen return period as described above, using the time of concentration (Section 2.7) as the duration of the storm. Some special cases of time of concentration are illustrated by the example given in the next section.

The area A is that of the roofed and paved areas drained to the pipe. Unlike the earlier method, however, not all the rain falling on these surfaces is taken as entering the pipes, nor is rain falling elsewhere on the catchment neglected. The volumetric coefficient, C_v allows for these complications. Values of C_v vary between 0.6 for areas of pervious soils to 0.9 for areas of heavy soils but the average, 0.75, applies quite widely as can be ascertained by applying the more detailed formulae for percentage run-off used in the hydrograph and simulation procedures.

The routing coefficient, C_R, allows for the combined effect of the disposition of impermeable area in the catchment and the variation of intensity during the storm. It was found to have a typical value of 1.30. This coefficient compensates for some of the shortcomings of the Rational method mentioned in Section 2.7.

2.11 Worked example of the Modified Rational method

Table 2.1 gives the calculations for the small surface water sewerage system whose layout is shown diagrammatically in Fig. 2.5.

Data applying to the whole calculation are given below the table. The values quoted for r and $M5$–60 min are to be regarded as having been ascertained for this particular site.

The chosen return period of one year would be adequate for a sloping residential area without cellars, where all the houses were at or above road level and where there were no other low spots where flooding might cause concern. Time of entry, t_e and z_2 follow from t as described in preceding sections.

Pipe roughness, k_s, is chosen according to the pipe material and is discussed in Chapter 6. If pipes of more than one roughness were to be used, an extra column would be provided in the table.

The first four columns of the table contain data which will be available at the start of the calculations. The impermeable area in column (2) is that of the roofed and paved surfaces drained through branches joining the particular pipe.

Column (3) gives the length of pipe between successive manholes.

The gradient in column (4) requires the longitudinal profile to be considered. Often the gradient will be the same as that of the ground surface to keep the pipe at a uniform, economic depth but in flat districts a steeper gradient may be needed to keep the mean running full velocity not less than 0.75 m/s or, with more prudence, 0.9 m/s. At this stage the profile of the pipe soffits is considered. Invert levels cannot be worked out until the diameters have been calculated, since changes in diameter are always accomplished by dropping the invert while keeping the soffit at the same level.

For pipes receiving flow from branches along their lengths, the diameter in column (5) is initially a trial value, because it will determine the velocity, time of concentration and, hence, the design flow. The smallest diameter used for sewers is 150 mm, nominal, and sizes increase in nominal steps of 75 mm.

In the example, velocity, column (6), has been calculated from

$$V = \left\{ \frac{2gds}{4f} \right\}^{\frac{1}{2}}$$

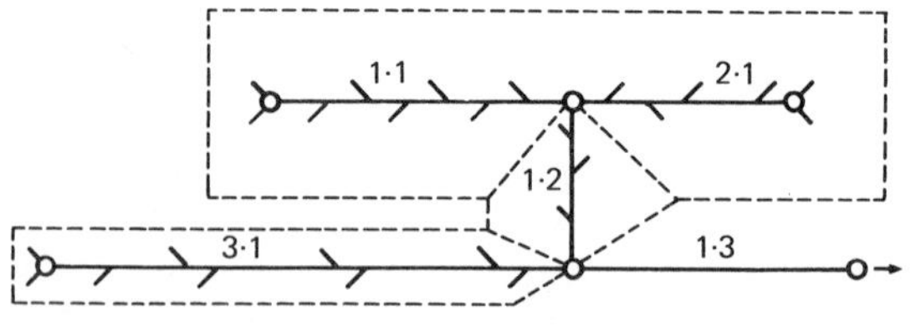

Fig. 2.5 Surface water sewers designed in Table 2.1

Table 2.1

Reference (1)	Imp. area ha (2)	Length m (3)	Gradient (4)	Diam. mm (5)	Velocity m/s (6)	Capacity litre/s (7)	t_f min (8)	t_c min (9)	Rainfall intensity I_5* mm/hr (10)	Total area ΣA (11)	Design flow litre/s (12)
1.1	0.125	95	1 in 71	150	1.05	18.6	1.51	7.51	72.92	0.125	14.83
2.1	0.093	70	1 in 100	150	0.88	15.6	1.33	7.33	73.69	0.093	11.17
1.2	0.210	52	1 in 182	225	0.86	34.2	1.01	8.52	68.44	0.428	47.66
1.2	0.210	52	1 in 182	300	1.03	72.8	0.84	8.35	69.15	0.428	48.15
3.1	0.015	164	1 in 100	150	0.88	15.6	3.11	9.11	66.10	0.015	1.61
1.3	0	90	1 in 154	300	1.13	79.9	1.32	9.11	66.10	0.443	47.64

$r = 0.39$
M5–60 min = 20 mm
$T = 1$ yr

$t_e = 6$ min
$z_2 = 0.61$
$k_s = 1.5$ mm

$C_v = 0.75$
$C_R = 1.30$
ARF = 0.985

Design flow $= C_v C_R z_2$ (ARF) I_5 (ΣA)/0.36 $= 1.627 \times I_5 \times \Sigma A$ litre/s.
M5 rainfall intensity, $I_5 = $ (M5–60 min) $\times (Z_1/D)$

in which $\ln\left(\dfrac{z_1}{D}\right) = \left\{ \dfrac{\ln(721/(11+15D))}{\ln(721/16)} - 1 \right\} \ln(48/r/1.06)$.

* for 5 yr return period; z_2, for conversion is included in the coefficient 1.627.

in which

$$\frac{1}{\sqrt{f}} = 4 \log_{10}(3.7 \ d/k_s)$$

These formulae are explained in Chapter 6. They are valid for $Vk_s/\nu \geqslant 807$, a condition usually met for the k_s values used for sewers.

Column (7), the pipe capacity is simply $V\pi d^2/4$, and column (8) the time of flow is column (3) divided by column (6).

Further explanation must proceed line by line as there are differences in arriving at the times of concentration and total areas. In the first line the time of concentration is the sum of the time of flow for the pipe and the time of entry. The 5-year rainfall intensity in column (10) is obtained by using the time of concentration as the duration of storm in equations (2.4) and (2.5).

The area in column (11) of the first line is the same as that in column (2). The design flow in column (12) is obtained from:

$$C_v C_R z_2 (\text{ARF}) I_5 \Sigma A / 0.36 = 1.627 \ I_5 \Sigma \ A \ \text{litre/s}$$

Since for pipe 1.1 the design flow does not exceed the capacity, the 150 mm diameter can be accepted.

The design of pipe 2.1 follows exactly the same procedure.

In the third line, for pipe 1.2, care is needed in assigning the time of concentration. Since this is the time taken from the furthest point on the catchment we must consider whether the longer route is via 1.1 or 1.2. Pipe 1.1 provides the longer time of flow so the time of concentration is made up of 6, 1.51 and 1.01 minutes, the time of entry and the times of flow in pipes 1.1 and 1.2 respectively. The total area, column (11), is the sum of the areas draining directly to pipes 1.1, 2.1 and 1.2.

For pipe 1.2 the design flow 47.66 litre/s is greater than the capacity of 34.0 litre/s so a new trial has to be made with the next larger diameter. This calculation appears on the fourth line.

Pipe 3.1, a new branch, is designed following the same procedure as pipes 1.1 and 2.1.

Pipe 1.3 differs from the others in that it has no branches bringing in additional flow along its length. It is designed by considering the flow at its *upper* end and since this does not depend on the time of flow along the pipe itself, trial is avoided and the columns of the table are used in a different order. Initially columns (5) to (8) are left blank and calculation begins with the time of concentration. Again this is the time from the furthest point in the catchment which in this case is via pipe 3.1, the route along 1.1 and 1.2 being shorter. Calculation of columns (10) to (12) proceeds as before and, having obtained the design flow, a diameter is selected which will convey it with the available gradient. This diameter is entered in column (5), after which columns (6) and (7) can be completed to demonstrate that the velocity and capacity are satisfactory. Time of flow in column (8) is needed only where there are further sewers downstream.

It must be emphasised that it would have been incorrect to have designed pipe 1.3 by taking the time of concentration down to its lower end. This would have led to a lower intensity of rainfall over an area no larger, thus giving a lower design flow. This lower flow is the one which the longer storm would give but the longer storm is not the one which gives the worst case.

The example illustrates a related feature of the application of the Rational method which occurs sometimes. Comparing the design flows for pipes 1.2 and 1.3 it will be seen that the larger one is that for pipe 1.2. Strictly, there-fore, this larger flow should be taken for pipe 1.3 since the storm causing it is evidently a worse case for this pipe. An even worse case can be envisaged when it is realised that the 8.35 minutes storm would produce flow in pipe 1.3 from part of the area drained by pipe 3.1. This could be estimated by taking a part of this area in proportion to the ratio of its individual time of concentra-tion to the duration of the storm: $0.015 \times 8.35/9.11$ which gives 0.014 ha. With the intensity for the 8.35 minute storm and the additional area, the design flow for pipe 1.3 becomes 49.73 litre/s.

Let us now suppose that pipe 1.3 has a small amount of area directly drained to it through branches along its length, say 0.015 ha. The design point now shifts to the lower end and the time of concentration is 10.43 minutes giving I_5 as 61.49 mm/hr. Coupled with the new total area of 0.456 ha this gives a flow of 45.82 litre/s, even lower than the original estimate. Again the flow is one which could occur, but the storm of 8.35 minutes lasting only long enough for the areas drained to pipes 1.1, 1.2, 2.1 and part of pipe 3.1 to contribute, would give a higher flow.

It was for cases like these that the area–time diagram and tangent methods were devised. Such methods are now said to be obsolete and they were certainly cumbersome. Their effect was to use a storm shorter than the normal time of concentration whenever significant parts of the time of con-centration were associated with parts of the area stretching out at either or both ends of the catchment. The shorter storm applied to the greater part of the area in the middle would give a higher flow.

These problems occur only rarely but with catchments of severely attenuated shapes, a few plausible combinations of parts of the area with shorter storms should be tried to see if a seriously higher design flow can be found.

2.12 Hydrograph methods

The Rational method aims at doing no more than estimating the peak flow to guide the choice of pipe diameter in routine design.

There are a number of circumstances where a hydrograph showing the varying rate of flow throughout a storm event is needed. These include the design of any features employing storage to modify downstream flow such as balancing ponds (Section 2.13) and storage storm overflows (Section 4.11). In existing systems, comparison of hydrographs calculated from particular storm events with hydrographs measured in the sewer during the same events enables evaluations to be made of the methods of calculation and data values to be investigated for the particular system so that future calculations can be more reliable.

The production of a hydrograph demands a realistic representation of the process of run-off generation from rainfall and therefore leads to a routine in which more confidence might be placed than in the simple Rational method. Development of the hydrograph methods thus led to a more reliable model which could be used to evaluate the Rational method and to replace it where necessary.

In the Wallingford hydrograph procedure,[2] rainfall is regarded as entering the sewer after flowing over surfaces which cause the flow to be modified by storage effects. These effects are taken into account by treating the surfaces as reservoirs in which flow, q, is related to depth on the surface, h, by $q \propto h^{3/2}$, and in which volume stored, S, is related to depth of storage directly: $S \propto h$, so that $S = k_r q^{2/3}$. The ordinates of the rainfall hydrograph are used in conjunction with this equation in a manner akin to that in Table 2.3 on page 29 to produce a hydrograph of the flow to the sewer. Study of data from the experimental areas enabled the routing constant k_r to be related to slope and area per gulley.

It was found that k_r was not so sensitive to catchment slope and to area per gulley for continuous variation of these factors to be needed so, for each of them, particular values representing three ranges are used in the program together with a further pair of values for sloping roofs, thus simplifying data acquisition and input as well as computation.

Allowance has to be made for the extent to which total run-off in a storm is less than the total rainfall over the area drained. Data from field observations again enabled equations to be developed. These relate percentage run-off to variables such as the percentage of impermeable area, the soil type and the 'urban catchment wetness index' which in turn is related to annual average rainfall.

For realism the total rainfall loss is taken as occurring partly as depression storage at the beginning of the storm and partly as infiltration and related losses uniformly distributed throughout the remainder of the storm. Depression storage was found to be related to slope for which the same categories could be used as for the routing constant k_r.

As implemented in the computer program, the storage routing procedure is applied to the rainfall intensity hyetograph after deducting depression storage, without taking into account at that stage the areas over which the rain has been falling. Since there is a matrix of three categories of area per gulley and three categories of slope, together with sloping roofs, there is a total of ten cases and the hyetograph is routed separately through all of them, the resulting ordinates being stored and re-called later for multiplying by the areas in each category which are calculated so as to be compatible with the estimated overall percentage run-off. The resulting hydrographs together with that for flow from upstream are then added together and form the input to calculations for the storage effects in the sewer pipe itself. For these a version of the well-established Muskingum method is used.

A standard shape of hyetograph is used corresponding in peakedness to the median found in summer storms.

The procedure is one solely for rainfall to run-off transformation. The effect of storms of different maximum depth according to duration is taken into account simply by repeating the whole calculation for storms of different durations, normally 15, 30, 60 and 120 minutes, to seek the hydrograph with the highest flow. The result was found to be relatively insensitive to storm duration so closer intervals seemed unnecessary.

Provision is made in the Wallingford procedure programs for taking into account sewer surcharge and also the effects of ancillary features such as storm overflows, detention tanks and pumping. Programs are available both for design and simulation. The cost optimisation of design taking into account depth and standard of construction of pipes was also looked into and procedures based on the Modified Rational method were evolved.

Use of the simulation program for investigating an existing system requires the monitoring of rainfall and the resulting flows at a number of carefully selected points for a period of five weeks or more. Hydrographs obtained from the sewer flow monitors, which record depth and velocity, are compared with hydrographs generated by the simulation program from the corresponding rainfall data. Poor agreement is then eliminated by seeking and correcting faults in the data relating to pipes and contributing areas. The computer model can then be used with standard severe storms to investigate needs for renovation or renewal.

2.13 Balancing ponds

When an area of previously open country is built upon, the peak flows in the streams increase due to the greater impermeability of roofs and pavings. Not only is more rain intercepted and prevented from infiltrating to become groundwater, but run-off is more immediate, times of concentration are reduced and storms of shorter duration with higher intensity become the ones which produce the highest flows.

In areas drained by relatively small streams, the surface water from newly built development can lead to a need for the enlargement of watercourses to prevent flooding. Where streams are culverted under railways and motorways, enlargement can be very costly.

Problems of this kind were particularly acute in some of the new towns and a solution adopted was to provide flood storage in the form of balancing ponds upstream of the points where capacity was restricted.

Balancing ponds may be on-stream or off-stream (Figs 2.6(a) and (b)).

An on-stream pond can be formed by placing some form of barrier across the watercourse and widening the channel upstream of the barrier. If more storage is needed the barrier can take the form of a dam creating an impounding reservoir in the valley. Some form of control device limits the

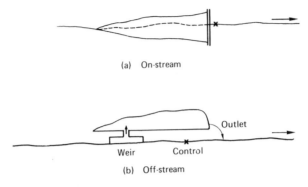

(a)   On-stream

Outlet

Weir   Control

(b)   Off-stream

Fig. 2.6   Balancing ponds

flow through the barrier. The control may be an orifice or a short length of pipe, a weir formed on top of the barrier, a gauging flume or some mechanically controlled gate. An overflow weir may be needed in addition to cope safely with floods of longer return period than those which the pond is normally intended to store. Separate consideration will have to be given to the downstream consequences of these. The Reservoirs Safety Act may be applicable.

Off-stream ponds are made alongside the stream, flow entering them by way of a side weir, siphon or penstock. Downstream of the side weir, a flume or other form of throttle limits the onward flow so that surplus spills into the balancing pond. At the end of a storm the pond empties back into the main channel through a pipe or orifice. If part of the storage volume is below the channel, pumps will be needed. In principle the off-stream pond has much in common with the storm tanks at a sewage treatment works.

A balancing pond may take the form of a permanent lake making a landscape feature, wildlife habitat or recreational facility. The storage volume is then provided above normal water level. If the lake is of sufficient area the water level will not rise very far during floods. Dry ponds, similar to flood washlands, offer another alternative.

The mode of operation of an on-stream pond is illustrated in Fig. 2.7. The rate of outflow depends on the head on the control device which in turn depends on the water level in the pond. Because of this the outflow hydrograph lags behind the inflow hydrograph and at any given time after the beginning of the hydrographs, more water has entered (the area under the inflow hydrograph) than has left (the area under the outflow hydrograph). The difference is the volume of water which is being held in temporary storage. The outflow hydrograph continues to rise until, towards the end of the storm, the inflow falls below it. From this point onwards, outflow is greater than inflow and the stored water gradually drains away.

For a given inflow hydrograph, the outflow hydrograph can be calculated (Section 2.14) using the relationships between stored volume, outflow and

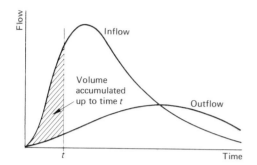

Fig. 2.7 Mode of operation of an on-stream balancing pond

water level. However, before undertaking this, a design storm has to be selected as a basis for the inflow hydrograph. Usually the severe storm needing the most storage is longer than the one giving the maximum instantaneous flow as used for pipe or channel design.

This can be shown as follows.[5,6] The volume of run-off in a particular storm may be expressed as $ApDI$ in which A is the area of the catchment, p the ratio of run-off to rainfall, D the storm duration and I its mean intensity, varying inversely with D. On the basis of the Rational model, the duration of this run-off is $(D + t_c)$ (Fig. 2.8). The volume to be stored is the difference in the areas abc and adc on the graph, of which abc is $ApDI$ and adc is $Q_m(D + t_c)/2$ if ad is approximated to a straight line or, say, $kQ_m(D + t_c)/2$ otherwise.

Thus the volume to be retained in storage is given by

$$R = ApDI - kQ_m(D + t_c)/2 \qquad (2.8)$$

The duration of storm which makes R a maximum can be obtained

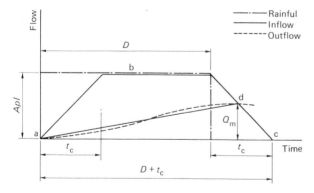

Fig. 2.8 Volume needed for a balancing pond on the basis of the Rational model. Q_m is the maximum downstream flow

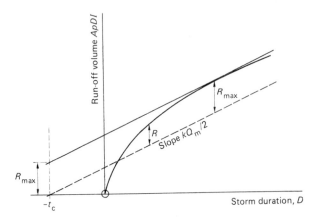

Fig. 2.9   Duration of storm for maximum storage requirement

graphically as shown in Fig. 2.9. The curve of $ApDI$ against D is obtained by using a relationship of I to D such as equation (2.4). The maximum value of R is the maximum distance between the curve and a line of slope $kQ_m/2$ starting from zero at $-t_c$. However, it is more convenient to draw a tangent to the curve with a slope of $kQ_m/2$ and to observe its intercept on the ordinate at $-t_c$. As can be seen, higher values of Q_m give reduced storage requirements, the duration at the point of tangency also being shorter. The value of D for zero storage is the one which the Rational principle would use for the instantaneous peak design flow.

In some cases it is the available storage volume which is fixed and the maximum downstream flow which has to be determined. The same graph can be used, the known storage volume being marked as the ordinate at $-t_c$ and the slope of the tangent from the ordinate giving the flow.

For a weir outlet a straight line approximation to ad in Fig. 2.8 is usually adequate. For outlets like orifices where head rises more rapidly with respect to discharge, a value of k greater than one may be needed. A full routing calculation (Section 2.14) can be done as a check once provisional design decisions have been reached.

The Wallingford hydrograph method can be used to provide a hydrograph for use in the routing calculations. Several storm durations should be tried to find the worst effect.

Storage to reduce onward flow can be used on a smaller scale than with balancing ponds by utilising fully the volume of the pipes themselves upstream of a control. Additional volume can be provided in separate off-stream pipes. A useful form of control for these purposes is the proprietary 'Hydro-Brake'. This throttles back high flows without obstructing low flows and works automatically without any moving parts. The vertical outlet model operates in a manner similar to that of the vortex drop (Section 7.8) but analogous phenomena of vortex motion and air-core flow above a certain discharge occur also when the axis is horizontal, and this is the more common form of Hydro-Brake. Several types are manufactured, usually in stainless steel, for installation in manholes.

2.14 Storage routing

The relationship between the inflow, P, outflow, Q, and volume, R, retained in temporary storage in a reservoir is given by

$$P = Q + dR/dt \tag{2.9}$$

However, R depends on Q since both R and Q depend upon the depth of storage in the reservoir above the outlet weir, or the equivalent for some other type of outlet.

For a weir outlet from a reservoir with vertical sides

$$Q = KLh^{3/2}$$
and $R = Ah$
so $R = A(Q/KL)^{3/2} \tag{2.10}$

K is a constant depending on the form of the weir.

The following step-by-step calculation enables P to be calculated from Q.

Using the notation of Fig. 2.10 to relate areas beneath and between the inflow and outflow hydrographs over a time interval Δt,

$$\frac{P_n + P_{n-1}}{2} \, \Delta t = \frac{Q_n + Q_{n-1}}{2} \, \Delta t + R_n - R_{n-1}$$

that is

$$P_n + P_{n-1} - Q_{n-1} + 2R_{n-1}/\Delta t = Q_n + 2R_n/\Delta t \tag{2.11}$$

To use this equation, a graph is first constructed of $Q + 2R/\Delta t$ against Q. Then at each step the left hand side of the equation is evaluated. This is

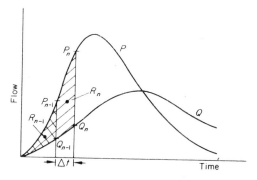

Fig. 2.10 The storage routing process

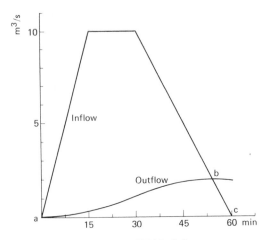

Fig. 2.11　Hydrographs of Table 2.3

possible because Q_{n-1} and $2R_{n-1}/\Delta t$ are available from the previous step. The value of the left hand side is then taken on to the graph to obtain the corresponding value of Q_n which can then be entered in the table and used to calculate $2R_n/\Delta t$ ready for the next step.

In the example shown in Figs 2.11 and 2.12 and in Tables 2.2 and 2.3, the peak outflow resulting from the input hydrograph shown is to be restricted to 2 m³/s by using a reservoir of 78 750 m² area with a weir outlet. The length of weir is to be determined and a routing calculation carried out to show that the weir length is suitable.

The maximum head on the weir must first be found from the storage

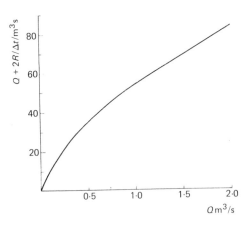

Fig. 2.12　Storage function relating Q + 2R/Δt to Q

Table 2.2

Q m^3/s	0.1	0.2	0.4	0.6	0.8	1.2	1.6	2.0
$Q + 350(Q/17)^{\frac{2}{3}}$	11.50	18.31	29.12	38.27	46.41	60.98	74.02	86.03

Table 2.3

n (1)	Time min (2)	P_n (3)	$P_n + P_{n-1}$ (4)	$Q_n + 2R_n/\Delta t$ (5)	Q_n (6)	$\dfrac{2R_n}{\Delta t}$ (7)
0	0	0	0	0	0	0
1	$7\frac{1}{2}$	5	5	5	0.03	4.97
2	15	10	15	19.94	0.22	19.72
3	$22\frac{1}{2}$	10	20	39.50	0.62	38.88
4	30	10	20	58.88	1.16	57.72
5	$37\frac{1}{2}$	$7\frac{1}{2}$	$17\frac{1}{2}$	74.06	1.60	72.46
6	45	5	$12\frac{1}{2}$	83.36	1.92	81.44
7	$52\frac{1}{2}$	$2\frac{1}{2}$	$7\frac{1}{2}$	87.02	2.04	84.98
8	60	0	$2\frac{1}{2}$	85.44	1.94	83.50
9	$67\frac{1}{2}$	0	0	81.56	1.84	79.72

volume needed. Approximating abc (Fig. 2.11) to a triangle, the area between the hydrographs gives a volume of

$$\tfrac{1}{2}(60 + 15) \times 60 \times 10 - \tfrac{1}{2} \times 60 \times 60 \times 2 = 18\ 900 \text{ m}^3$$

This is the maximum volume to be stored above weir level and with a surface area of 78 750 m^2 it will result in a head of 18 900/78 750 = 0.24 m. Assuming a weir coefficient of 1.7, the required length of weir is given by:

$$2 = 1.7\ L\ (0.24)^{3/2} \text{ m}^3/\text{s}$$

from which L is 10 m.

If a time interval of $7\frac{1}{2}$ minute (450 s) is to be used for the routing calculation:

$$Q + \frac{2R}{\Delta t} = Q + \frac{2 \times 78\ 750}{450}\ h$$
$$= Q + 350(Q/17)^{2/3}$$

since $\quad h = (Q/1.7L)^{2/3} \quad$ for the weir.

This relationship, needed in the step calculation, Table 2.3, is shown in Table 2.2 and Fig. 2.12.

In Table 2.3, columns (2) and (3) give details of the input hydrograph. Sums of successive pairs of ordinates are entered in column (4). At the nth step the value in column (5) is obtained from equation (2.11) by adding the

$(n-1)$th value from column (7) to the nth value from column (4) and subtracting the $(n-1)$th value from column (6). The result is taken on to Fig. 2.12 to obtain Q_n for column (6) which in turn is used by subtraction from column (5) to give the entry for column (5).

References

1 British Standards Institution CP 2005: 1968, *Sewerage.*
2 Department of the Environment, National Water Council, *Design and Analysis of urban storm drainage, the Wallingford procedure*, 4 vols, NWC, London 1981.
3 Shakespeare, W. *King Richard the Second*, Act II, scene I.
4 Institute of Hydrology, *Flood Studies Report*, 5 vols, Natural Environment Research Council, 1975.
5 Davis, L.H., The Hydraulic Design of Balancing Tanks and River Storage Ponds, *The Chartered Municipal Engineer*, **90**, 1–7, Jan. 1963.
6 Davis, L.H. and Woods, D.R., Design and Construction of Balancing Lakes at Milton Keynes, *The Chartered Municipal Engineer*, **106**, 9–17, Jan. 1979.

3

Constructional aspects of sewer design

3.1 Structural aspects of buried pipes

Buried pipes can fail in a number of different ways and from a number of different causes. These may be summarised as follows[1, 2].

1 General overload in the vertical plane causing longitudinal fracture at springing level, followed sometimes by fracture along the crown and invert.
2 Local overload due to hard spots beneath the pipe causing diagonal tension cracks to spread from the point of overload.
3 Relative movement of successive lengths of pipe causing circumferential fracture due to either excessive shear or excessive bending stresses. Such fractures can occur in the pipe barrel or in the socket. The relative movement causing such fractures may be due to non-uniformity of support, ground movements or differential settlement of a manhole or other structure into which the pipe is built.
4 Direct longitudinal tension due to thermal or drying shrinkage of the pipe or of concrete associated with it or to drying shrinkage of a clay soil or to the tension phase of mining subsidence. The fracture is circumferential and may be in barrel or socket.
5 Direct longitudinal compression due to restraint of thermal or moisture expansion or to the compression phase of mining subsidence. The actual fractures are longitudinal, due to circumferential tension and are usually at the joints.
6 Excessive pressure inside the joint due to expansion of the jointing material causing a bursting failure of the socket.

A particular failure may occur when a number of causes combine to produce excessive stress at a point where there is already a weakness due to faulty manufacture, storage or handling.

Failure of type 3 can be avoided by using flexible joints[3] provided that relative angular movement is not excessive. Flexible joints for clayware pipes allow deflections up to 5°. The deflections for concrete pipes depend on diameter and are somewhat less. If the limit is exceeded, the socket will be broken by leverage.

If a flexible joint allows telescopic movement, as many do, failures of types

4 and 5 are avoided as well as those of type 3, though type 5 failures are avoided only if a margin for expansion is left when making the joint. It is not unknown for flexible joints to suffer socket bursting failures (type 6) but the hazard is very much less than with cement mortar joints.

Though less skill is needed for flexible joints than for cement mortar joints, good supervision is still needed to ensure that the work is done carefully, particularly with large diameter pipes. Pressure testing after laying provides an essential incentive to good practice. The length of pipeline to be tested is plugged at the downstream end and has a vertical length of pipe attached to the upstream end by means of a bend or a blanked tee. The pipe is filled with water so that the head on the pipe is nowhere less than 1.2 m, nor greater than 6 m. After initial loss during the first hour has been made up, topping-up water to replace further losses is added from a measuring can at three 10 minute intervals and should not exceed half a litre per metre length of pipe per metre of pipe diameter, according to CP 2005. An alternative air test, allowing 25 mm water gauge of drop from 100 mm in five minutes, avoids the use of water but is no more than a screening test and would not alone give sufficient ground for rejection.

3.2 Matching pipe strength to imposed load

From the preceding section it may be concluded that the occurrence of all modes of failure except the first can be minimised by careful construction and the adoption of telescopic-flexible joints. In order to minimise the occurrence of the first mode of failure it is necessary to estimate the maximum load which is likely to be imposed on the pipe in the vertical plane and then to choose a pipe which, with the addition of some form of support or protection where necessary, will be strong enough to withstand the imposed load.

The components of imposed load are as follows.

1 The effect of the earth above the pipe, which may include an embankment to be tipped later above the original ground level.
2 An imaginary external downward load which is equivalent in terms of pipe-wall stress to the effect of the water contained in the pipe.
3 The effect of loads superimposed on the surface of the ground. Super-imposed loads may be classified as follows.
(a) Uniformly distributed surcharge of wide (assumed infinite) extent taken as the equivalent of any surcharge which might occur and used in place of surcharge loads estimated in detail as in 3(c) and 3(d) below where appropriate.
(b) Uniformly distributed surcharge of small extent to cover permanent surcharges from building foundations.
(c) Uniformly distributed surcharge of small extent and of transient nature arising from heaps of earth or building materials.
(d) Point load surcharge from vehicle wheel loads with an allowance for impact effects.

Though earth load increases with depth of cover, the effect of wheel loads decreases with depth. Consequently, the highest loads which pipes have to sustain are at shallow depths (due to wheel loads) or at large depths (due to earth load). At intermediate depths the imposed loads are lower. In many circumstances pipes can be laid without external protection at these intermediate depths.

Before methods of calculation were introduced in the U.K. designers were guided by a set of rules laid down for loan sanction purposes, applying to clayware and concrete pipes up to 900 mm in diameter. For diameters up to 450 mm with depths of cover between 1.2 m under roads or 0.9 m not under roads and 4.3 m, no protection was required. For diameters between 450 mm and 750 mm at these depths the pipes were to be bedded and haunched with 150 mm of concrete. At cover depths between 4.3 and 6.1 m all pipes up to 750 mm were to be bedded and haunched. The haunching was carried up to half pipe depth and finished with 45° slopes tangential with the pipe wall. At cover depths greater than 6.1 m or less than 1.2 m (roads) or 0.9 m (fields) pipes had to be completely surrounded with 150 mm of concrete.

Because of the complexity of the methods of calculating loads these rules continued in use but tables and graphs are now available applying to the most commonly encountered conditions. Though more extensive than the old rules, the tables are no more difficult to use. Furthermore, they relate to methods of construction which are considered to be preferable.

The succeeding sections outline the theory upon which pipe load calculations are based. Tables and graphs[5, 6] available in most design offices avoid the need for calculations from first principles but do not explain fundamental bases. The designer needs to understand these to appreciate the significance of assumptions made and to identify and analyse correctly the special cases which are occasionally encountered.

Estimates of loads in categories (1) and (3a) above are based on a theory of soil behaviour, involving several simplifying assumptions, originally propounded by Marston eighty years ago and further developed by Marston, Schlick and Spangler at Iowa State University.

The methods of dealing with loads in the other categories were also developed by these workers but in categories (3b), (3c) and (3d), Boussinesq's analysis of pressure within a semi-infinite elastic medium due to load at the surface is used. The Boussinesq theory is not in any way combined with the Marston theory; loads from the distinctly separate theoretical approaches are simply added together. It would anyway be extremely difficult to reconcile the two approaches. The main justification for present practice is that such field data as are available are not markedly in conflict with the results of calculation.

3.3 Loads due to earth overburden

Several loading cases have been distinguished.

(1) Narrow-trench case

This applies when the trench is relatively narrow and relatively deep in relation to the width of the pipe or conduit structure. There is an analogy with pressure at the bottom of a silo.

Notation and the basis of analysis are shown in Fig. 3.1. The object is to obtain the total load per unit length of pipe on the horizontal plane X-X extending to the trench sides. This is taken as the load to be carried by the pipe on the assumption that a form of arching action will occur below X-X. Some allowance for this assumption is involved later in the application of bedding factors to assess pipe strength. In German practice, the fill adjacent to the pipe below X-X is considered to carry some of the load but this involves an assumption as to the standard of workmanship in placing the fill.

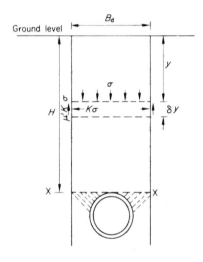

Fig. 3.1   Notation for derivation of narrow-trench coefficient

The settlement of the trench fill due to the deformation of the pipe will mobilise friction stress between the fill and the trench sides. This stress will act upwards on the fill, reducing the total load.

Analysis is performed by considering the equilibrium of an element, δy, of fill-depth. Vertical stress σ is assumed to be uniform across the element.

Rankine's formula is used for horizontal component of stress:

$$[(1 - \sin \phi)/(1 + \sin \phi)] \, \sigma = K \, \sigma$$

in which ϕ is the angle of internal friction of the fill material. Strictly this applies only to granular materials, but for simplicity it is used indiscriminately.

The friction stress is $\mu' K \phi$ in which μ' is the coefficient of friction between the fill and the trench sides.

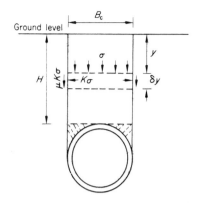

Fig. 3.2 Notation for derivation of wide-trench coefficient

For equilibrium of vertical forces per unit length on the element:

$$B_d \sigma + \gamma B_d . \delta y - 2 \mu' K(\sigma + \delta\sigma/2) \delta y - B_d (\sigma + \delta\sigma) = 0$$

in which γ = weight of unit volume of fill.

Integrating this equation betwen limits of $y = 0$ and $y = H$ gives:

$$\sigma_H = \gamma H \{1 - \exp(-\alpha'_d)\} \alpha'_d$$

in which $\alpha'_d = 2 \mu' K(H/B_d)$.

Load on the plane X-X per unit length of pipe = $\sigma_H B_d$.

It is customary to express this as $C_d \gamma B_d^2$ in which

$$C_d = \frac{1 - \exp(-\alpha'_d)}{2 \mu' K} \tag{3.1}$$

The expression is shown graphically as curve (a) in Fig. 3.3 in the form α'_d versus $2 \mu' KC_d$ so as to be independent of the value of $\mu' K$.

(2) Complete wide-trench or embankment case

If the trench sides are remote from the pipe as in a wide-trench, or absent as in a pipe laid on the original ground with an embankment tipped on top of it, the load upon the pipe is greater than the weight of fill immediately above it and on the Iowa theory it is considered that vertical shear planes, tangential to the pipe sides, will form as shown in Fig. 3.2. The additional load arises from friction stress on these planes mobilised by settlement outside the shear planes in excess of that within the planes which is restricted by the presence of the pipe.

Analysis similar to the above but with friction stress reversed leads to:

$$C_c = \frac{\exp(\alpha_c) + 1}{2 \mu K}, \alpha_c = 2 \mu K (H/B_c) \tag{3.2}$$

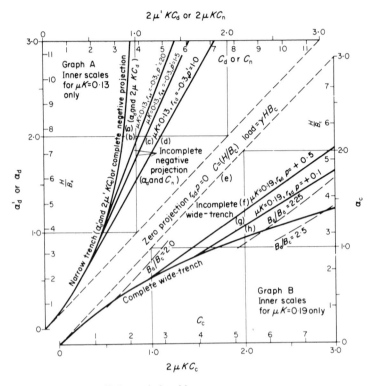

Fig. 3.3 Load coefficient relationships

load per unit length being given by $C_c \gamma B_c^2$. Here, μ is the coefficient of internal friction of the fill.

The relationship between $2 \mu K C_c$ and α_c is shown as curve (h) in Fig. 3.3.

(3) Incomplete wide-trench case

The description of the preceding case implies that, when settlement occurs, the fill level above the pipe will be slightly higher than the fill level alongside. If the depth of overburden is increased, the pipe deformation will increase and the extent to which the fill over the pipe protrudes above the surrounding fill will decrease until at a certain height there will be no protrusion and shear at the ground surface will be zero. If further fill is added the shear planes will not extend into it and indeed the 'plane of equal settlement' at which the shear planes die out will fall slightly as the total depth of overburden increases. This case is shown in Fig. 3.4.

The height of the plane of equal settlement above the top of the conduit, H_e, depends on the projection of the pipe above its foundation, h, and on the deflection of the pipe relative to the settlement of fill alongside over the

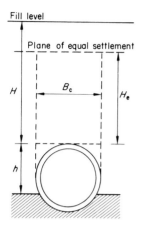

Fig. 3.4 Incomplete wide-trench or embankment case. For an unyielding foundation, as shown $p = h/B_c$. Values commonly accepted for $r_{sd}p$ are 1.0 where bedding is Class A or B, 0.5 where bedding is granular

depth h. This is measured by r_{sd}, the settlement-deflection ratio. The development of r_{sd} is shown in Fig. 3.5.

The left-hand half of this figure defines a 'critical plane' in relation to the hypothetical case when there is neither settlement of the fill nor deformation of the pipe. The right-hand half takes account of settlement and deformation. Between the left-hand and right-hand halves, the following changes have occurred: the depth of fill h has decreased by s_m due to

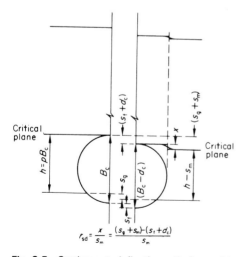

Fig. 3.5 Settlement–deflection ratio for positive projection

settlement, the foundation alongside the pipe has yielded by an amount s_f, and the pipe diameter has deformed or deflected by an amount d_c. Thus the critical plane has dropped to the extent $(s_g + s_m)$ and the top of the conduit to the extent $(s_f + d_c)$. The protuberance x of the top of the pipe above the critical plane is therefore $(s_g + s_m) - (s_f + d_c)$. The ratio $x/s_m = r_{sd}$ is taken as the measure of the relative settlement and deflection. The product $r_{sd}p$ in which $p = h/B_c$ controls the extent of the shear planes and the height of the plane of equal settlement. To cover different forms of bedding, $r_{sd}p$ is often regarded as single entity.

The derivation of the load-factor expression for this case is straight-forward differing from that for the *complete* case, 2 above, only in that the integration is performed over the interval H_e, the load at the plane of equal settlement being $\gamma(H - H_e) B_c$. The result is:

$$2 \mu KC_c = \exp(\alpha_{ec}) - 1 + (\alpha_e - \alpha_{ec}) \exp(\alpha_{ec}) \qquad (3.3)$$

in which $\alpha_{ec} = 2 \mu K (H_e/B_c)$.

Since H_e depends upon H as well as upon μK and $r_{sd}p$, this expression is not adequate alone. The equation proposed by Spangler for obtaining H_e is extremely complex and has to be solved by trial but the combined effect of the two equations is to produce a nearly linear relationship between C_c and α_c as shown in lines (f) and (g) in Fig. 3.3.

The wide-trench case follows the curve of equation (3.2) until H/B_c is large enough for the *incomplete* case to be encountered and then follows the straight-line approximation appropriate to the values of μK and $r_{sd}p$. The remainder of curve (h) is irrelevant. The linear expression may be used for the whole relationship with little error (*complete* case loads being slightly overestimated)

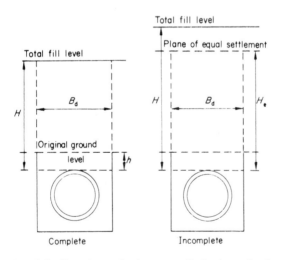

Fig. 3.6   Negative projection cases. Projection ratio $p' = h/B_d$

amounting to a constant percentage increase on the weight of overburden above the pipe, $\gamma B_c H$.

(4) Negative-projection cases

This is an extension of the narrow-trench case and applies only where an embankment is placed above a pipe which is in a trench below original ground level. There is a similarity to the wide-trench case in that a plane of equal settlement occurs with large total depths of fill so that both complete and incomplete cases arise (see Fig. 3.6).

Figure 3.7 shows how r_{sd} is defined for *negative* projection. The ratio r_{sd} has a negative sign and the datum surface from which h' is measured is the original ground level, instead of the pipe foundation. It is because h' is measured downwards from datum level that the term negative projection is adopted (the wide-trench case may be termed *positive* projection). It should be noted that $r_{sd} p'$ is negative but the negative sign arises from the manner in which r_{sd} is defined, p' and h' being taken as positive.

The load-factor expressions for the negative-projection cases are:

Complete case: $2 \mu K C_n = 1 - \exp(-\alpha_d)$,
$$\alpha_d = 2 \mu K (H/B_d) \tag{3.4}$$

Incomplete case: $2 \mu K C_n = 1 - \exp(-\alpha_{ed}) + (\alpha_d - \alpha_{ed}) \exp(-\alpha_{ed})$,
$$\alpha_{ed} = 2 \mu K (H_e/B_d) \tag{3.5}$$

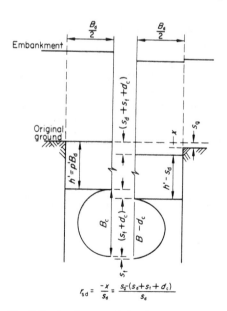

$$r_{sd} = \frac{-x}{s_d} = \frac{s_g - (s_o + s_t + d_c)}{s_d}$$

Fig. 3.7 Settlement–deflection ratio for negative projection

A complex equation connects H_c with H, μK and $r_{sd}p$ but the incomplete case again gives a nearly linear relationship, lines (*b*) and (*c*) Fig. 3.3.

(5) Zero-projection case

This is the limiting case where positive projection or $r_{sd}p$ is zero and may be considered to occur where the top of the pipe is near original ground level with an embankment as overburden. No shear planes are considered to form and the load is simply the weight of overburden immediately above the pipe, γHB_c, that is, $C = (H/B_c)$.

In some circumstances, it may be arguable that the load should be taken over the trench width, as γHB_d or $C = (H/B_d)$. Zero projection is then the limiting case for negative projection.

(6) Negative r_{sd} in wide trenches

With pipes made from flexible material, it is possible for the top of the pipe to be depressed below the critical plane so that r_{sd} becomes negative. The fill above the pipe then moves downwards in relation to the fill alongside and the shear stress is reversed. This leads to expressions algebraically similar to those of the negative-projection cases except that B_c replaces B_d:

$$\text{Complete: } 2 \mu KC_c = 1 - \exp(-\alpha_c)$$
$$\text{Incomplete: } \quad 2 KC_c = 1 - \exp(-\alpha_{ec}) + (\alpha_c + \alpha_{ec}) \exp(-\alpha_{ec})$$
$$\alpha_c = 2 \mu K(H/B_c)$$
$$\alpha_{ec} = 2 \mu K(H_e/B_c)$$

The incomplete case may again be approximated to by linear relationships[4,6] but these are different from those for negative projection. Though $r_{sd}p$ is negative both for the present case and for negative projection, the meanings of r_{sd} are not truly analogous and the equations for determining H_e differ, as also do the straight-line approximations to the relationship of C to α.

Pipes of material to which this subsection applies are encountered rarely in sewerage work and data as to bedding factors, wheel-load effects, etc., are at present insufficient to enable them to be designed reliably. Their strength is probably more than adequate in most circumstances.

(7) Values of μK and $r_{sd}p$

Values of μK commonly quoted are given in Table 3.1. It is common where

Table 3.1 Values of μK and $\mu' K$

Granular, cohesionless materials	0.19
Maximum for sand and gravel	0.165
Maximum for saturated top soil	0.150
Ordinary maximum for clay	0.130
Maximum for saturated clay	0.110

detailed information of the nature of the soil is lacking to use a high value of μK for the wide-trench cases and a low value for the narrow-trench and negative-projection cases in order to err on the safe side. Thus 0.19 is often used with C_c and 0.13 with C_d and C_n.

Commonly used values of $r_{sd}p$ are 1.0 where the pipe bedding is rigid and does not extend to the full trench width, and 0.5 for beddings such as Class B (see Fig. 3.9) or concrete-arch beddings.

In negative projection $r_{sd} = -0.3$ has been tentatively adopted for rigid pipes.

(8) Transition between narrow-trench and wide-trench cases

For trenches which are deep or narrow in relation to conduit breadth the narrow-trench case will be appropriate. At some relatively small depth, the two cases will give the same load value and, for still smaller relative depth, the wide-trench case will apply as the trench sides will be relatively too far from the pipe to exercise their influence. The general rule is that the case to be used is the one which gives the lower load value.

Figure 3.3 will be used to illustrate the occurrence of the transition. It is first necessary to convert the α-values into H/B_d and H/B_c for specific values of μK. In accordance with common practice, $\mu K = 0.13$ is used for graph A and $\mu K = 0.19$ for graph B, so that any error in estimating μK will be on the safe side. The H/B scales are given on the insides of the vertical axis. A scale of C_c is added to the horizontal axis of graph B and a scale of C_d or C_n to the horizontal axis of graph A.

Adopting a particular value for B_d/B_c, curve (a) can be replotted in relation to the H/B_c and C_c scales of graph B according to the relationships:

$$\frac{H}{B_c} = \frac{H}{B_d}\left(\frac{B_d}{B_c}\right)$$

and
$$C_c = C_d \left(\frac{B_d}{B_c}\right)^2$$

The second relationship arises through the need to make total load equal $C_c \gamma B_c^2$ for either case.

Curves for several ratios of B_d/B_c obtained in this way are shown on graph B. The relationship of load to H/B_c as depth of cover increases may now be seen to follow curve (h) and possibly line (f), or (g), until the curve for the appropriate value of B_d/B_c is encountered, after which the B_d/B_c curve (for narrow-trench loading) is followed.

For a given value of B_d, the value of H at which transition occurs is known as transition depth, while for a given value of H the value of B_d at which transition occurs is known as transition width. Whenever narrow-trench loading is taken it is important to ensure that B_d does not exceed the assumed design value.

Where there is to be an embankment above original ground level, the negative-projection relationship — curve (a) breaking off along a line such as

(b) — would be used to produce the B_d/B_c curves on graph B. It will be apparent that, with increasing depth, several different sequences of cases are possible according to the values of B_d/B_c, $r_{sd}p$ and $r_{sd}p'$.

Loading cases are brought together for comparison in Fig. 3.8.

(9) Effect of water-table above pipe

Below water-table level the unit weight of soil is reduced by the effect of buoyancy and the fill load as calculated from saturated but unsubmerged soil will be reduced by an amount $C_{ds} \gamma_w B_c^2$ according to the loading case. In these expressions γ_w is the unit weight of water and the load coefficients are similar to C_d and C_c but with α'_{sd} and α_{sc} replacing α'_d and α_c. The α factors are:

$$\alpha'_{sd} = 2 \mu' K (H_s/B_d), \quad \alpha_{sc} = 2 \mu K (H_s/B_c)$$

Since the effect of a high water-table is to reduce the load, submergence should not be taken into account unless it is permanent. The possibility of future changes in water-table level can seldom be excluded and it is rare that submergence can be safely considered in design.

3.4 Internal water load

Water within the pipe imposes stress on the pipe wall and this is allowed for by adding to the external vertical load an amount which will cause stress equivalent to the effect of the water. For circular pipes this amount is approximately three-quarters of the weight of water in a full pipe, varying with bedding factor, F_m, and of little significance for pipes smaller than 600 mm. For non-circular sections either a stress analysis must be attempted (for large, reinforced concrete sections this would be implicit in the design calculations) or the weight of water in the full section must be taken so as to be on the safe side.

Where the fluid in the pipe is under pressure, stresses additional to those caused by the fill-loading will be created. The maximum internal pressure may be transient or temporary due to water-hammer or testing but may nevertheless have to be taken into account. The possibility of temporary internal vacuum may also need to be considered.

3.5 Uniformly distributed surcharge of wide extent

If such a surcharge of weight per unit area, U, is considered to be of infinite extent, the effect will be to increase all pressures beneath it to the extent U. It is convenient to express U as an equivalent depth of additional fill,

$$\Delta H = U/\gamma$$

Analysis is performed as for the fill loads but taking $\sigma = U$ at $y = 0$. The resulting expressions for the total of fill plus surcharge in the form $2 \mu K C$ are

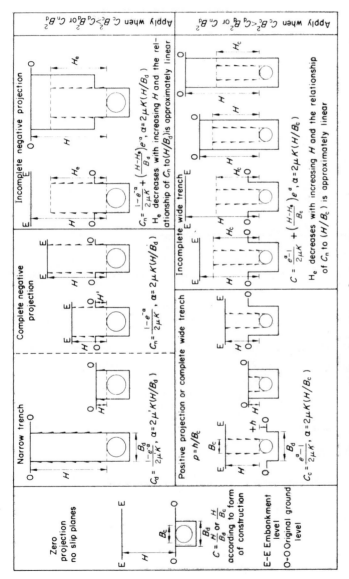

Fig. 3.8 Loading cases for buried pipelines (*After* Walton[7]). The half arrows on the slip planes indicate the direction of friction force on the fill supported by the pipe

Table 3.2   Total of fill load and effect of uniformly distributed surcharge of wide extent (The formulae are expressions for load factor in the form $2\,\mu KC$ or $2\mu'KC$. $\Delta H = U/\gamma$ in which U is the surcharge per unit area.)

Narrow trench	Zero projection
$1 - \exp(-\alpha'_d) + \Delta\,\alpha'_d\exp(-\alpha'_d)$ $\alpha'_d = 2\mu'K(H/B_d)$ $\Delta\,\alpha'_d = 2\mu'K(\Delta H/B_d)$ <div align="right">(3.6)</div>	$C = (H + \Delta H)/B$ Load $= \gamma\beta(H + \Delta H)$ <div align="right">(3.9)</div>

Negative projection	Wide trench
Complete	
$1 - \exp(-\alpha_d) - \Delta\alpha_d\exp(-\alpha_d)$ $\alpha_d - 2\mu K(H/B_d)$ $\Delta\alpha_d = 2\mu K(\Delta H/B_d)$ <div align="right">(3.7)</div>	$\exp(\alpha_c) - 1 + \Delta\alpha_c\exp(\alpha_c)$ $\alpha_c = 2\mu K(H/B_c)$ $\Delta\alpha_c = 2\mu K(\Delta H/B_c)$ <div align="right">(3.10)</div>
Incomplete	
$1 - \exp(-\alpha_{ed}) +$ $(\Delta\alpha + \alpha_d - \alpha_{ed})\exp(-\alpha_{ed})$ $\quad\quad\alpha_{ed} = 2\mu K(H_e/B_d)$ $\Delta\alpha_d + \alpha_d - \alpha_{ed} = 2\mu K(\Delta H +$ $H - H_e)/B_d$ <div align="right">(3.8)</div>	$\exp(\alpha_{ec}) - 1 +$ $(\Delta\alpha_c + \alpha_c - \alpha_{ec})\exp(\alpha_{ec})$ $\quad\quad\alpha_{ec} = 2\mu K(H_e/B_c)$ $\Delta\alpha_c + \alpha_c - \alpha_{ec} = 2\mu K(\Delta H +$ $H - H_e)/B_c$ <div align="right">(3.11)</div>

given in Table 3.2. The formulae for the incomplete cases are similar to those for the same cases without surcharge and the linear expressions for the latter can be used with the simple modification of substituting $(H + U/\gamma)$ for H. Note that this implies that the addition of the surcharge lowers the plane of equal settlement.

In deriving the equations in Table 3.2, the surcharge has been assumed to have no internal friction. The presence or absence of friction within the surcharge is irrelevant in the incomplete cases.

If the surcharge has the same internal friction as the fill, the complete wide-trench load factor for fill plus surcharge becomes:

$$2\,\mu\,KC = \exp(\alpha_c + \Delta\,\alpha_c) - 1 \tag{3.12}$$

instead of the expression (3.10) in Table 3.2.

Though surcharge is taken as having no internal friction for narrow-trench and negative-projection cases, friction is taken as acting within the surcharge for the wide-trench case and (3.12) is customarily used instead of (3.10). The justification for this inconsistency is that it errs on the safe side and produces a simpler formula. With this simplification $(H + \Delta H)/B$ may replace H/B for both wide-trench cases.

In calculating surcharge effects it is usual to consider the case of a narrow trench or wide trench according to which is critical for the fill load alone. In other words, transition depth is assumed to be unaltered by the addition of surcharge. This results in the total load decreasing abruptly when depth changes from just below to just above fill-load transition depth. The lower values are not, however, taken for design. Instead the wide-trench value obtained at fill-load transition depth is used for greater depths until a depth is reached at which it is exceeded by the narrow-trench load.

More strictly according to the theory, transition depth is reduced by the addition of surcharge, and the narrow-trench case is encountered without a reduction in load. The practice of assuming that transition depth is unaffected by surcharge results in overestimation of the load over the range of depth from fill-with-surcharge transition depth to the depth at which the narrow-trench case is commonly allowed to prevail.

Since the effect of extensive surcharge is considered for comparison with other surcharge loads it is necessary either to estimate it separately from fill load or to subtract fill load from the total of fill and surcharge when this is obtained in a single operation. The value normally taken for U is 23 kN/m^2. The effect of this is added only when it is less than the effect of concentrated surcharges such as wheel loads. This is rarely the case. It has been recommended[3] that extensive surcharge need not be considered, since uniform surcharge is unlikely to exceed 23 kN/m^2 over an area 4.5×4.5 m in extent and the effect of this is less than that of recommended wheel loads. When uniform surcharges of greater finite extent are expected they should of course be considered.

3.6 Loads superimposed by vehicle wheels

The theoretical basis for estimating these effects is Boussinesq's formula, which enables the vertical component of stress to be calculated at any point below the surface of an elastic solid, extending infinitely from the surface, due to a point load upon the surface. When it is desired to find the total vertical load on a horizontal area beneath the surface, stress has to be integrated over the area. Methods of doing this were developed by Holl, Newmark ('influence values') and Fadum. This work is dealt with in textbooks on soil mechanics and will not be repeated here. Clark and Young[8] used it to obtain effective load values resulting from vehicle wheels to suit conditions in the United Kingdom.

The process of calculation consists of several stages, some of them requiring extensive computation so it is not practicable to calculate individual cases. Instead the graphs in the TRRL *Guide*[5] are used in routine design. A brief outline of the development of effective load values will be given.

With the train of wheels in a chosen position, the loads from the several wheels over a horizontal area tangential to the top of the pipe are assessed using Newmark's influence values. Since the effect of a point load extends indefinitely along the pipe, becoming smaller as distance from the load increases, a specific length of pipeline must be chosen over which the load is to be calculated. A length of 1 m is now used since it has been shown that the choice of length is not critical.

Account has to be taken of the non-uniformity of stress over the area upon which load is calculated. The ultimate object is to produce a load value which can be added to the fill loads. The load value is therefore taken as the uniform load which will give the same pipe stress as the calculated non-uniform load. Factors have to be worked out to enable the loads obtained from the Boussinesq-Newmark stage to be converted to equivalent uniform loads. The variation of soil stress over the area and consequently the conversion factors depend on the depth of cover, on the breadth of the pipe and on its form of support and bedding.

In given circumstances there will be a critical position of the wheel train which gives the greatest load on the pipe. The axles may be at right-angles or parallel to the pipe axis. The distribution of soil stress over the area at the top of the pipe differs according to the position of the wheels. The position giving the worst effect depends again on depth and breadth of pipe and on the form of bedding.

The graphs in the TRRL *Guide*[5] give unit pressures, σ, to be multiplied by conduit width, B_c, to give uniformly distributed load, W_{csu}, equivalent to the effect of wheel loads:

$$W_{csu} = \sigma B_c$$

Curves are provided for main roads, lightly trafficked roads, fields, and for single and multiple track railway. Impact factors have been allowed for in calculating the unit pressures. Advice is given on allowances to be made for construction vehicle loads.

3.7 Uniformly distributed surcharges of limited extent

Surcharges of this nature allow for temporary heaps of material on the ground surface or for the effects of foundations of buildings or other structures. Except where they are not directly above the pipeline they will replace rather than supplement wheel loads.

The effects of these loads are allowed for by using influence values to assess the total load from the surcharge over a horizontal plane at the top of the pipe similar to that for vehicle wheel loads.

3.8 Pipe strength

A pair of diametrically opposed line loads is the uniquely simple loading condition and is chosen as the basis for pipe-strength considerations. For pipe materials which possess a high degree of uniformity it would be feasible to calculate safe loads for this condition, using safe values of stress obtained by testing samples of the pipe material. For materials such as clayware and concrete, which are likely to vary significantly in quality, it is necessary to test whole pipes to find failure loads directly. It is not always practicable to test pipes under perfect line-loading conditions. Practicable testing methods may require slight departures from the perfect line-loading condition and involve some spreading of the load so that test values are higher than those for line loading. A conversion factor C_{BS} is applied so that

Equivalent line load $= W_T\, C_{BS}$ per unit length of pipe

in which W_T is the load applied in the testing apparatus. For the tests commonly used for clayware and concrete pipes, C_{BS} is taken as about 0.95 for pipes larger than 825 mm diameter and 0.90 for smaller pipes. In choosing a working safe value of W_T, the variation of test results is taken into account so that only an acceptably small percentage would fail at loads less than the chosen value.

As laid in the field, pipes will sustain greater vertical loads even than in standard laboratory test conditions as the fill surrounding the pipes distributes the load round the barrel. This is taken into account by expressing 'field strength' as

$W_T\, C_{BS}\, F_m$ per unit length of pipe

The bedding factor, F_m, depends on the form of construction. It is the ratio of the vertical load applied by the fill to the line load which would cause the same maximum circumferential bending stress. For pipes laid under embankments or in very wide trenches, different bedding factors, F_p, apply. These are higher than trench-bedding factors for corresponding forms of construction since side thrust on the pipe reduces the maximum circumferential bending stress.

The type of pipe and form of bedding is chosen so that

$$W_T\, C_{BS}\, F \not< F_s\, W_e$$

in which $F = F_m$ or F_p as appropriate,

W_e = total downward effective load due to fill and superimposed loading, calculated as described in previous sections,

and F_s = load factor to provide a safety margin.

The factor C_{BS} is omitted[3] as its application is considered to be an insignificant refinement especially as it is closer to 1.0 with modern testing methods.

Two forms of pipe strength, W_T are in use, the proof strength, $W_{T(proof)}$ and the ultimate strength $W_{T(ult)}$. The former is the load per unit length sustainable under test without a crack of more than specified size developing. The latter is the load per unit length sustainable without collapse irrespective of crack width.

With $W_{T(proof)}$, F is taken as $\not< 1.0$ while with $W_{T(ult)}$ it is taken as 1.25. Both requirements must be satisfied so for pipes with $W_{T(proof)} < 0.8\, W_{T(ult)}$, proof strength must be used with F_s as 1.0.

It should be noted that W_e itself depends on the form of bedding or protection. Thus in the incomplete cases, choice of bedding affects $r_{sd}p$ and hence the formula needed for C_c or C_n. The effects of vehicle wheel loads differ for different bedding classes. The use of concrete arch protection affects the estimated applied load in several ways. Cover depth is reduced, decreasing fill load but increasing the effect of vehicle wheel loads. Conduit width B_c is increased to the width of the concrete affecting both H/B_c and $C_c \gamma B_c^2$.

It is the operation of factors such as these added to the interrelation of the fill-load cases which renders it difficult to produce comprehensive charts and tables which would avoid calculation. Though it is possible to tabulate data for the circumstances most frequently encountered, which have less complexity than those of rarer occurrence, there is always the danger that they may be used unwittingly for circumstances to which they should not be applied.

Fill loads increase with depth while wheel-load effects decrease with depth. The combination of the two causes total load to decrease with increasing cover depth until a minimum is reached, after which total load increases with increased depth. A consequence of this is that total load for a given length of pipeline has to be assessed for both the least and the greatest cover depth along the length to discover which condition gives the greater load.

Trench bedding factors, F_m, for some forms of construction are shown in Fig. 3.9. Class 'A' offers the highest strength but is not often needed. Prior to the calculation of loads, pipes over 450 mm in diameter or with more than 4.3 m of cover were required to be laid with 150 mm of concrete beneath the pipe and carried up to mid-depth, so many existing sewers will be found to have this form of protection. Where cover depths were less than 1.2 m under loads or 0.9 m elsewhere, complete concrete surrounds were provided and this is still commonly done where pipes have small amounts of cover, particularly in building drainage. Complete surrounds were also provided where cover exceeded 6.1 m.

An alternative class 'A' construction to the concrete cradle shown in Fig. 3.9 is to place a concrete arch above the pipe but this has rarely been adopted except as a means of increasing the strength of a pipe after it has been laid, because it makes side connections more difficult.

Class 'S' bedding has the pipe completely surrounded with granular material for which detailed specifications have been issued[9]. A similar type of bedding is known as class 'E' in Scotland.

The granular material has usually to be imported to the site, sometimes from a considerable distance. Nevertheless bedding of this class is not uncommonly specified, particularly for smaller pipes, since it is more straightforward to use the same material for both the surround and the cover for the pipe. Where granular bedding is used it is sometimes advisable to provide water stops at intervals along the trench.

Class 'B' bedding uses granular material only up to half pipe depth and so is cheaper but offers a correspondingly lower bedding factor. It is perhaps the commonest form of construction. Above the granular bedding, to a depth of

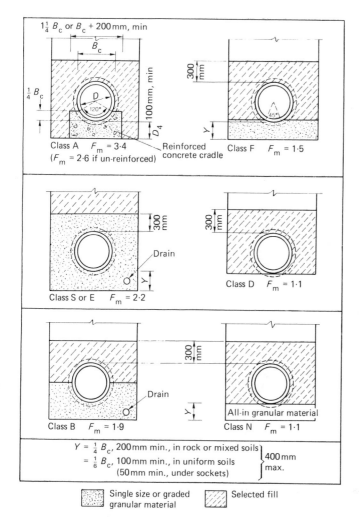

Fig. 3.9 Some pipe trench beddings. For detailed specifications and other forms of bedding see ref. 6

300 mm above the top of the pipe, selected fill is used. This is to be material which is readily compacted, free from vegetable matter, rubbish, frozen soil, clay lumps larger than 75 mm and stone larger than 25 mm. Often it can be selected from the excavated material.

Class 'D' and the alternatives to it, 'F' and 'N', are often referred to as 'minimum beddings'. In class 'D' the trench bottom has to be carefully trimmed by hand to support the pipe barrel, holes being made for the sockets. This method cannot be used in rock nor in very soft clays or silts, nor even in very compact sands and gravels. In these materials, class 'N' construction is used. Here a 'regulating layer' is provided underneath the pipe.

Another alternative where the ground is unsuitable for class 'D' is class 'F' which, both in construction and in bedding factors, is intermediate between classes 'D' and 'B'. It is little used, probably because its particular form places too much reliance on careful workmanship.

In all forms of bedding, compaction of fill at the sides of the pipe to give proper lateral support is of great importance.

The increasing cost of granular bedding material coupled with improved pipe manufacture and the availability of stronger pipes points towards wider use of minimum beddings especially as there is a belief that the bedding factors in current use may err on the safe side. Under the old rules, pipes up to 450 mm in diameter and with cover between 1.2 and 4.3 m were laid in a manner not dissimilar to minimum bedding. Many miles of these remain in service, though it was because of instances of failure that the Marston theory was introduced into British practice.

Forms of bedding where an embankment is to be placed above the pipe are given in the TRRL *Guide*[5] as well as detailed specifications and design graphs.

3.9 Constructional materials for sewers

Brief details of pipe and other materials used for sewers are given below with notes on their typical ranges of application. Up-to-date British Standards and manufacturers' catalogues need to be consulted for precise details of diameters, lengths and joints, as changes are made, and availability and prices vary, from time to time. In some cases the price depends on the location of the site, so the best choice may differ from one scheme to another.

(1) Clayware pipes

In diameters up to 225 mm, clayware is the most commonly chosen material for sewers and in many places clayware pipes are used up to 450 mm diameter. The older sewers have cement mortar joints but these were superseded some years ago by rubber ring joints, which are flexible and telescopic. In the smaller diameters socketless pipes with polypropylene, push-fit couplings are common, especially in building drainage work. Methods of manufacture which ensure that the clay is vitrified through the full thickness of the pipe wall have obviated the need for salt glazing.

Extra-strength pipes are available for use where loads are high, or to enable the cheaper forms of bedding to be used. Pipes are made also for thrust-boring (Section 3.13) and slip-lining (Section 3.15). There is a full range of bends and junctions including half-pipes to form manhole channels.

(2) Concrete pipes

Socketted concrete pipes with flexible rubber ring joints are made in diameters from 0.15 to 2.7 m. Ogee jointed pipes are made in diameters up to 1.8 m and a type of rubber ring, flexible joint for them is now available. Diameters increase in steps of 75 mm up to 1.2 m diameter and in steps of 150 mm thereafter. Three strength classes are provided.

Pipes made with sulphate resisting cement instead of ordinary Portland cement should be used wherever ground water is likely to be sulphate bearing even if the pipes are to be surrounded with sulphate resisting concrete.

Pipes with thicker walls are made for pipe-jacking operations, special pipes being used behind the shield and at intermediate jacking stations.

Porous concrete pipes, 0.1 to 0.9 m in diameter, are useful for land drainage.

(3) Brickwork

There are many brick sewers in existence and they were formerly the usual choice for diameters of about 0.9 m and above but have been restricted to successively larger diameters as the size of concrete pipes has increased. Brickwork is still used as the lining for large reinforced concrete sewers constructed *in situ* (Section 3.12) where aggressive conditions are expected.

(4) Cast-iron pipes

These are described as ductile spun iron pipes, the material differing from the grey spun iron formerly used by the presence of spheroidal rather than lammelar graphite. Pipes up to 2 m in diameter are available in the U.K. Socketted straight pipes are 5.5 m in length.

Class of pipe, K (for 'Klass'), is related to wall thickness, e, by

$$e = K(0.5 + 0.001d) \text{ mm}$$

in which d is the nominal internal diameter in mm. For standard socketted pipes K is 9 while for flanged pipes and fittings other than tees, K is 12. For tees it is 14. Test and working pressures are specified according to diameter, being lower for the larger diameters. They are given in BS 4772 and ISO 2531.

Cast-iron pipes are normally coated inside and out with some form of bitumen compound but can be supplied with spun cement mortar linings.

For pipelines in trench and on bridges, spigot-and-socket pipes are used. Flexible joints such as the bolted gland and the 'Tyton' are more common than run or caulked lead joints. The bolted gland joint is sealed by a rubber ring held in place by a loose flange bolted to a flange formed round the socket collar. The 'Tyton' joint has a socket which is grooved to retain the rubber sealing ring once the spigot is driven in.

Where pipes are exposed in pumping stations, etc., flanged joints are used. Though the pipes are held together quite positively, it is prudent to support them from the structure so that stresses due to their own weight and that of their contents are minimised. Wherever flanged pipework is used, the layout must be such that the joints can be properly tightened. The presence of bends makes this possible in many instances, but a flexible-telescopic joint such as a Johnson coupling (see below under 'steel pipes') is sometimes needed in addition.

The application of cast-iron pipes in sewerage is as follows: where high crushing strength is required and pipes of other materials with special protection would be more expensive or would take too long to lay; where sewage is under pressure as in pumping mains and in inverted siphons; where a very high degree of erosion resistance is needed; where a pipe is above ground on piers or on a pipe bridge.

There is a separate class of cast-iron pipe for drains together with a wide range of special fittings. These pipes are used where heavy loads are likely to occur, as in factory yards, and wherever a drainage system has to be carried out to the very highest standards. Spun pipes are used for the long straight lengths in such work. Some of the fittings of the drainage range can usefully simplify the constructional details of certain special manholes and treatment works units.

(5) Steel pipes

When used for ordinary sewers or pumping mains, steel pipes are most commonly plain-ended and are joined with Johnson couplings, which are in effect double bolted glands in a separate collar.

Where a steel pipe is to span a gap without support, long-sleeved welded joints or some other joint with bending strength is needed.

Steel pipes are about 8 m long but are not manufactured to precise standard lengths. The range of diameter and wall thickness is very wide and will cover any conceivable pipe-sewer application.

Adequate corrosion protection is very important. There are several forms of factory-applied internal and external sheathing and coating. Any damage done to the coating must be made good on site and protection must be completed at each joint after laying by pouring bitumen into a mould surrounding the coupling. A liner enables the inside of the joint to be protected by bitumen flowing through a hole in the coupling.

Cathodic protection is sometimes adopted to prevent corrosion.

Steel pipes have applications similar to those of cast-iron pipes. Since they are lighter, longer and ductile there are situations where steel is preferable to cast-iron but it is always necessary to protect steel pipes thoroughly from corrosion if they are to have a comparable life. In instances where shock loads are likely to occur either internally (water hammer) or externally, steel pipes should be chosen, but steps should be taken to see that the possibility of an internal vacuum is avoided.

(6) Asbestos-cement pipes

The two varieties of asbestos-cement pipe of interest here are pipes made especially for sewerage purposes, and pressure pipes made primarily for water supply as an alternative to cast-iron. The former are an alternative to clayware and concrete pipes. Pressure pipes may be used for rising mains in place of cast-iron or as gravity sewers where extra strength is needed. These pipes are long and have plain ends which are joined by forcing into separate collars. With suitable equipment, transport and handling over fields are simplified and there are fewer joints to make. They are sometimes used for long runs in rural areas. They are easily damaged and care is needed both in handling the pipes and in backfilling the trenches. Bitumen coating is needed if ground water is likely to be acidic or sulphate-bearing.

(7) uPVC (unplasticised poly-vinyl chloride) pipes

These are made in diameters up to 600 mm and in wall thicknesses furnishing four strength classes. They have socketted joints for rubber rings or for solvent cement and are in lengths of 6 to 9 m.

They are light, flexible and are highly resistant to both chemical attack and erosion. However, they lose strength as temperature rises and become brittle at freezing temperatures, and with age.

3.10 Excavation of trenches for pipe sewers

Whenever possible, sewer trenches are excavated with a mechanical excavator rigged as a back-acting shovel. The machine travels backwards with its tracks straddling the line. Spoil is placed well away from the side of the trench. Depth is gauged by means of a boning rod and sight rails, see Fig. 3.10. There is a sight rail at each end of the line, set at a predetermined level, which is the length of the boning rod above the invert level. Sight rails crossing the trench as shown are used for setting pipes to correct level and for guiding hand excavation but for machine excavation they have to be clear of the trench on alternate sides.

Trench supports are fixed as the excavation proceeds. Commonly used supports consist of corrugated, perforated steel sheets held apart with timber or adjustable steel props. If the ground is poor or if the trench is close to buildings, continuous sheeting is needed and excavation has then to be carried out by hand, though an excavator will usually be used as a crane to lift skips of excavated material. Accidents to personnel due to the falling in of trench sides where insufficiently supported have been regrettably common. Trench shoring machines remove much of the risk.

To facilitate drainage of the trench, excavation proceeds in the upstream direction. If there is a lot of water, a channel may have to be formed at the side of the trench bottom and sumps off the main line of the trench may be needed for pumping.

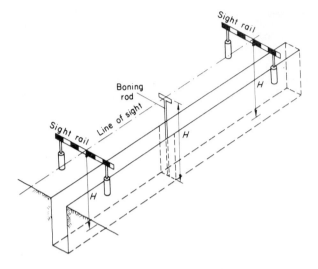

Fig. 3.10   Use of sight rails and boning rod

Exceptionally, the ground has to be de-watered before excavation by a well-pointing system alongside the trench. Freezing has had to be resorted to occasionally. Such methods increase the expense of construction very considerably and in any area where there is a possibility of their being needed a thorough site investigation is justified at the design stage to see whether a cheaper line can be found. It is arguable that site investigations should be carried out and the results furnished to tenderers (with necessary safeguards) more generally.

If the trench formation is liable to soften up on exposure it is 'blinded' with weak concrete immediately after excavation. A rich mix must not be used or pipe fractures may occur.

When the work is in the highway it is advisable to prepare a scheme before tenders are invited to show which lengths must be completed and temporarily reinstated before further lengths are opened. Such a scheme, even if it has to be modified, saves a good deal of argument during construction and settling up.

3.11 Sewers above ground

Because it is desirable to concentrate sewage treatment at a few points rather than to have many small plants, the area drained to a particular plant does not always coincide with a natural catchment area. Sometimes in these cases it is necessary to run a sewer line across an area of low-lying land. Often a longer route could be chosen such that the sewer would remain below ground level. The additional cost of the longer route may have to be balanced against the visual disadvantage of having a pipeline on piers above ground level.

Shorter lengths of above-ground sewers occur where railways, rivers and canals have to be crossed.

Where pipes may have to be above ground level, an inverted siphon or a pumping system can be considered as an alternative but this would usually be more costly and less satisfactory in performance.

Cast-iron and steel pipes are strong enough to span between piers and do not need continuous support. Since steel pipes need stringent protection against corrosion, cast-iron pipes are more commonly used. Steel pipes will span wider gaps and may be preferable where it is advantageous to have fewer piers, as for a high pipeline.

Where cast-iron is used the pipes usually have spigot-and-socket joints. They are supported by pier caps in the form of concrete cradles shaped to the pipe barrel. The piers are spaced one pipe length apart, the pipes being laid so that there is a cradle just behind each socket. Temporary supports are needed during construction. Sheet lead or bitumenised felt is used to spread pressure evenly between the pipe-barrel and the cradle.

Joints of the bolted-gland type are now commonly preferred. These avoid the high bending stresses which would be induced by differential settlement of piers if more rigid joints were used. If care is taken to leave gaps in the joints, no special expansion joints are needed.

For spans equal to the pipe length, steel pipes may be jointed with Johnson couplings. However, depending on wall thickness and diameter, steel pipes will span up to 10 or 12 m. Long sleeve-welded joints must then be used to transmit bending stresses. In areas subject to mining subsidence, flexible joints could still be used at the piers but it would usually be better to use flexible joints throughout with a pier at each joint to avoid the need for special short lengths. Where there is no risk of subsidence a steel pipeline with long spans is designed as a continuous beam with rigid joints. Provision for expansion and contraction will then be needed. Special expansion joints can be used at long intervals (perhaps at the ends of the pipeline) or Johnson couplings at shorter intervals. When a Johnson coupling or other expansion joint is used, a pair of closely spaced piers is provided to support both sides of the joint.

Continuous pipes with rigid joints are supported on rollers shaped to the profile of the pipe barrel. The piers are topped with bearing plates on which the rollers travel. On a long run, the pipeline is anchored on piers half-way between the expansion joints. At the anchor piers, the pipe is held by an arrangement of steel straps bolted together.

Design data and details for long steel pipelines are available from pipe manufacturers who collaborate on the design. The rollers, plates and supports, are supplied with pipes.

Lower piers are constructed in brickwork. It may be desirable to provide an outer skin of facing bricks. Reinforced concrete would be used for high piers. Structures of rolled steel sections will often be cheaper for high piers, even though they need painting periodically, but are rather unsightly.

Structures of tubular or hollow sections are preferable in appearance. The possibility of lateral forces should not be forgotten.

A well-proportioned design for the pipeline and its supports need not cost more than one which is ill-proportioned. The help of an architect should be sought at an early stage.

Where the pipeline is no more than a few feet above ground it may be concealed by an embankment. Unless adequate consolidation of the pipe bed can be achieved, piers should still be provided.

A bridge may be needed where a river, railway or canal has to be crossed. For a small-diameter pipe this would be a lattice structure of rolled, tubular or hollow steel sections. Roller supports for the pipes are provided at panel points. In detailing the bridge attention must be paid to access for repainting.

Where the pipeline is not too small in diameter, a self-supporting steel pipe can often be used and is neater in appearance than a bridge. For longer spans, composite structures on the suspension bridge principle have been used for water and gas mains, the deck consisting of the pipe which is supported by cables. The changes in profile caused by expansion and contraction render them less suitable for free-surface flow sewers but there may be cases where they would be feasible.

Large sewers may be designed to span gaps as reinforced (possibly prestressed) concrete box-sections. The main trunk sewer of the Bolton and District Joint Sewerage Board crosses the River Irwell by means of bridges of bow-string arch form, the sewer forming the chord member.

3.12 Large sewers

It is rare for pipes of over 2 m diameter to be used, though pipes up to 2.5 m diameter are not unknown. Where greater capacity is needed some form of *in situ* conduit must be constructed. Depending on circumstances an *in situ* form of construction can be cheaper than pipes even for sections less than 2 m in diameter. The length needed of a given section, the time during which the trench may be open and the accessibility of the site are all important factors.

Though many large brick sewers have been built to a circular cross-section, there is a free choice of shape with *in situ* construction. Donkin[10] showed many years ago that a U-shaped section with a flat roof had advantages over a circular section. Less width is needed for a given flow and low-flow velocities are slightly better. A good reason for using circular sections is that the whole section can act as a mass brick or concrete arch whereas construction must be partly if not wholly of reinforced concrete if the section is not wholly curved. This does not necessarily make circular sections cheaper.

Flat-topped sections have, in theory, the characteristic that they can accommodate more flow when the top is just not wet than when they are running just full. For a section of equal breadth and depth, the discharge is about 1.3 Q_o (see Section 6.5) just before the top is wetted and the practice adopted for circular sections of regarding Q_o, the running full discharge, as the design flow would be unduly conservative for flat-topped sections. It is better to regard the roof as being unwetted and to ensure that this will be so by having the roof safely above the intended maximum depth of flow. The margin is a matter of judgement depending on the likely accuracy of the estimates of peak flow and of channel roughness.

An alternative to the true U-shaped section with circular invert is to have a level invert with curved corners. Sections of this type have been used for sewers carrying up to 6 × DWF where low-flow characteristics are not particularly important.

If the minimum flow is a very small fraction of the peak flow, a 'Paris' section with a dry-weather flow channel below the general floor level is worth considering, Fig. 6.7.

It can be shown that the carrying capacity of a section of given shape is mainly dependent on the area. The effect of varying the breadth-to-depth ratio within practicable limits is trifling. This property is useful when an optimum breadth-to-depth ratio is being sought. A section with depth greater than breadth is usually cheapest, as excavation is reduced without greatly increasing other costs.

True U-sections which are not too large can be constructed in mass concrete and brickwork, only the roof needing reinforcement. Sections which are nearer to a box-shape need to be reinforced throughout and it is usually economic to build the larger U-sections in reinforced concrete, especially if the roof is monolithic with the walls. There is a considerable structural advantage, whatever the section, in making a monolithic box. Mid-span moments in the roof are reduced and the walls act as partially fixed-sided slabs rather than as cantilevers.

In carrying out the reinforced concrete design, BS 5337: 1976 should be used. Several loading conditions have to be considered. It is unarguable that the sewer must be able to withstand earth and superimposed loads when it is empty as well as when it is full. Controversy sometimes arises as to whether it should be designed to withstand internal water load without external earth loading. Circumstances can easily be envisaged where this loading condition could occur even if only locally and it should normally be taken into account.

The values of maximum bending moments are required at corners of the section and at mid-span of walls, floor and roof. Owing to the variety of loading conditions, both positive and negative moments can occur at most of these points. All likely combinations of load (including live load on the ground surface) must be considered and maximum moments, both positive and negative, sought at each point. Steel will be needed at both faces of the slabs almost everywhere as moments of either sign will be possible.

Longitudinal reinforcement must not be neglected. It is needed to resist shrinkage stresses rather than as 'distribution' steel and also to aid the fixing of the main reinforcement.

Whether expansion, or more strictly contraction, joints should be provided at intervals along the length is a matter for argument but they are a wise insurance. If they are omitted a generous provision of longitudinal steel should be made.

Large foul sewers are usually lined with brickwork. If the sewage is expected to be particularly aggressive, due for example to trade waste, first quality engineering bricks are used, at least for the lower part of the lining. The bricks are jointed with Portland cement mortar. The expense of special mortars is not often felt to be justified except for sewers specifically intended for strong trade waste, especially as no universally resistant material is available. The possibility of deterioration of joints must be faced. Regular inspection and maintenance are essential as areas of lining can break up quite rapidly if a defect occurs. Because frequent attention is so important, manholes should not be skimped either in number or for ease of use.

3.13 Sewers constructed underground

Tunnelling or some other form of underground construction may be warranted in the following circumstances.

1 To make a route more direct by tunnelling under a ridge.
2 To make a route more direct by tunnelling under buildings.
3 To avoid disturbance of road or rail traffic.
4 Where the depth of the sewer and the nature of the overburden are such that open-cut construction would be more expensive.

Once it becomes evident that a scheme is likely to involve some tunnelling, the aim is to produce an overall plan which achieves the best combination of open-cut and underground construction. The nature of the soils and rocks will be even more important than in a scheme which is to be entirely in open cut.

Thorough site investigation will be essential but a preliminary scheme based on geological maps and local knowledge will have to be drawn up first to enable boreholes to be sited. Fairly large changes in line may have to be made and further boreholes sunk where underground conditions are complex.

Long lengths of tunnel require shafts spaced along the route. It is desirable to minimise the time each working shaft is in use so that plant can be fully employed. Tunnelling is therefore carried out in both directions simultaneously from the foot of the shaft and an attempt is made to space the shafts in such a way that the drives on each side of a shaft are about equal in length. Drives up to about 400 m are feasible on sewers of about 2 m diameter,[11] the length of drive being roughly proportional to the diameter. The need for longer drives, which would require artificial ventilation, seldom arises in sewer tunnels as manholes have to be provided at fairly close intervals. A manhole is built in each working shaft after tunnelling is completed. Usually there are a few manholes in addition to those in working shafts and the excavations for these can be used to provide additional ventilation if long drives are desirable.

Sewers up to about 1.2 m diameter can be constructed from pipes laid in a timbered heading. For small pipes the dimensions of the heading will be determined by working space requirements, while for larger pipes the dimensions will be chosen to give a specified minimum thickness of concrete packing around the pipe. If reinforced concrete side trees are used for the heading and left in, they may be included in the cover thickness. In good ground the top of the heading may be supported by curved steel plates, reducing the volume of excavation and of concrete packing.

With timbered headings there is a danger that in poor ground, timber will have to be left in. This will minimise immediate settlement but may lead to settlement later when the timber rots. These problems together with the expense of such highly skilled work and the availability of newer techniques such as thrust-boring, have made timber heading a much rarer expedient.

Where the internal diameter is to be 1.4 m or more, the sewer may be built up from internally flanged pre-cast concrete or cast-iron segments, assembled in rings inside the tunnel and bolted together as excavation proceeds. Holes in the segment webs enable grouting to be carried out. The segments are lined with concrete and sometimes with an inner lining of brickwork.

Where high ground stresses have to resisted, cast-iron segments are preferred to those of concrete. Spheroidal graphited cast-iron has superseded grey cast-iron as it allows sections to be thinner and has some resistence to tensile stress.[12]

Interlocking concrete segments are a cheaper alternative to bolted segments. These have no flanges and are of uniform thickness (about 0.1 m) so that the inner surface has no panels to be packed with concrete. During construction, the precast units are held together in rings by erector plates which have dowels fitting into holes in the units. The erector plates are removed later. The tunnel is given an inner lining of 20 mm of cement mortar applied in two layers by gun or by a centrifugal lining machine. Brick lining may be used if foul sewage is to be carried, particularly if trade wastes are present.

Segmental construction, bolted or interlocking, is a soft-ground technique. In rock, tunnels may be driven with little support. They may be lined with pipes, packed and grouted externally, or with *in situ* concrete and brickwork, placed after the excavation has been completed.

In bad ground it may be necessary to drive a segmented tunnel by means of a shield. This consists of a steel sleeve fitting round the outside of segment rings already laid. The upper part of the shell is extended, forming a hood to protect the working face. As excavation proceeds, the shell is advanced by hydraulic jacks bearing on the last ring of segments. When the shield has advanced far enough, the jacks are retracted and a new ring is placed. Bolted segments are considered preferable for shield-driven tunnels.

If the water-table is high, sub-drains leading to pumps at the shaft may be inadequate for drainage and compressed air working may have to be resorted

Fig. 3.11   Thrust boring (*Courtesy Rees Construction Services Ltd*)

to. Alternatives may be ground freezing, chemical injection or well-point de-watering. Specialist advice and a specialist contractor are needed. Tunnelling is not undertaken in these circumstances unless it is unavoidable and an important object of site investigation is to discover whether difficulties needing such techniques are likely to arise.

Thrust boring (see Fig. 3.11) may be cheaper than traditional tunnelling for relatively short lengths. The risk of disturbing structures on the surface is lessened so that in suitable ground a thrust bore can be used even if the cover is not very great. The principle is shown in Fig. 3.12. As excavation proceeds at the working face, the pipes are thrust along by jacks in the shaft. As the jacks reach the limit of travel they are retracted and packings are inserted successively until the line has been advanced a whole pipe length. The packings are then removed and the next pipe placed in the bottom of the shaft. Small adjustments can be made to the direction of travel, to maintain line and level, and high standards of alignment can be achieved.

Long drives are possible if carried out in stages. After about half the pipes have been pushed in, a steel sleeve is added with a second set of jacks. The next pipe enters the sleeve with the jacks retracted. The original set of pipes is then thrust forward by the sleeve jacks which are then retracted so that the shaft jacks can push the following pipes along. The two sets of pipes are

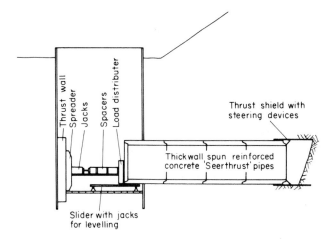

Fig. 3.12 Thrust boring (*Courtesy Rees Construction Services Ltd*)

pushed along alternately until the full length is in place. The sleeve has to be left in.

Thrust boring has been used mainly where busy roads or railways have to be crossed. Conduits up to the size of a pedestrian subway have been successfully installed by this technique. Further information can be obtained from the Pipe Jacking Association of specialist contractors.

When small-diameter pipes are thrust-bored, an auger is used to perform the excavation. Pipes or conduits of smaller size than those used for sewers can be thrust with a solid nose-cone which forces a passage through the ground, no material being removed.

The development of the proprietary 'Mini Tunnel' system (Mini Tunnels International, Old Woking, Surrey) which is free from difficulties sometimes experienced with thrust boring has enabled tunnelling to be used for small sewers and at smaller depths than formerly. It can compete with open cut construction along busy roads.

3.14 Sewers under water

Where a sewer has to be laid in a river bed the traditional method is to construct cofferdams. Since the whole of the waterway cannot be obstructed at once an area of bed extending to just beyond the middle of the stream is enclosed first and kept dry by pumping. Pipes are laid within the enclosure and the dam is then removed, a second cofferdam being constructed from the opposite bank to enable the line to be completed.

If the river is wide and cannot be crossed in two stages, intermediate cofferdams are used and a bridge, ropeway or boat is needed for access.

Where the sewer is fairly small so that construction can be quickly completed and the river is shallow, a primitive muck fill or sand-bag dam may be sufficient. For deep rivers and elaborate construction, interlocking steel sheet piles are used and a trestle will be needed to form a driving platform.

Small streams are most easily dealt with by diverting them completely through a temporary channel, the natural channel being completely dammed both upstream and downstream of the working area.

It is rarely reasonable to make a cofferdam so high that the risk of its being overtopped is negligible. Where a substantial proportion of the waterway is to be obstructed, serious flooding may be caused upstream if a high flow occurs. It may be better to keep the dam fairly low and to risk the possibility of having to repair the works rather than to risk damage elsewhere.

Because cofferdam construction is relatively expensive, it is becoming more common to use techniques which enable pipes to be laid under water. A trench is dredged across the river bed and the pipe is dropped into it by one of a variety of methods.[13] Concrete can be placed to refill the trench by tremie pipe (a vertical pipe with a hinged flap at the bottom) but sometimes silting can be relied upon to refill the trench. With a narrow river, cranes on the shore can be used to place the pipeline. With a wide river the pipe can be floated out and lowered. It can be supported by floats or floated in a sealed condition and sunk by admitting water. Another alternative is to use a lay barge from which the pipeline can be lowered. In many instances, the whole length of pipeline can be handled at once by one of these methods, the pipeline having been assembled on shore. Steel or PVC pipes are used because a measure of flexibility is essential. If the pipeline has to be laid in sections a diver is needed to complete the jointing.

Techniques used for laying pipelines out to sea, developed for the oil industry, are another possibility. The pipes are laid out side by side on the shore parallel to the direction of the proposed pipeline. The first pipe is pulled out over rollers into the water until the second pipe can be moved into line and jointed to it. After a further pull, the third pipe is attached and so on until the complete pipeline is assembled. There is a choice between pulling the pipe along the bottom and pulling it along the water surface. A bottom pull can be used only where there is little friction between the pipe and the bed. With a surface pull, the pipeline is supported by buoyancy chambers which are attached to the pipeline at suitable intervals as pulling proceeds. The pipeline is allowed to settle by admitting water to the chambers when it is on the correct line.

Specialist contractors are needed particularly for pulling techniques as a good deal of skill and experience is required. The choice of technique depends on the height and steepness of the banks, the nature of the bed and on whether a navigation passage has to be maintained.

3.15 Sewer renovation

With the ageing of sewerage systems, the renovation of existing sewers is becoming as important as the laying of new ones and recent years have seen the development of techniques and equipment which avoid excavation and the disruption of traffic which results from work in the highway. Even though excavation and reinstatement costs are greatly reduced by such methods factors such as the use of expensive materials and the large amount of pains-taking preparatory work make these techniques quite costly though not quite so costly as full replacement. Though the durability of most of the new materials seems unquestionable it has to be admitted that there is little long term experience of them.

The most obvious method of renovation is slip-lining in which a slightly smaller pipe is pushed or pulled through the existing one. High and medium density polyethylene are commonly used materials for the pipes but special clayware pipes have been developed also. By adding the pipes in short lengths as the work proceeds, only small excavations are needed at the ends of the pipelines and existing manholes might provide enough space.

With some forms of lining, holes at lateral connections are cut in the lining before it is inserted, careful measurements having been made in the sewer with the aid of closed circuit television.

Slip linings usually need grouting into the existing pipe with a material such as polyurethane foam. Where pipe defects are of small extent, localised repairs can be carried out by injection of resin around a rubber packer expanded within the pipe to maintain the cross-section. Repair of this kind may be all that is needed. In other cases local repairs may be needed before re-lining the whole length. Ensuring that voids around the sewer are filled is one of the most important as well as the most difficult aspects of renovation.

Any form of lining within an existing pipe reduces the diameter but may not reduce the carrying capacity to the same extent since the lining material is usually smoother than the original pipe and in many cases a number of obstructions to the flow will have had to be removed in order to insert the lining.

Pipes can be replaced without the loss of diameter by the use of a bursting mole towed through the old pipe. The leading end of the mole is conical and has fins which, with the aid of high frequency hammer action, split the pipe into four or more sectors which are pushed into the surrounding ground as the mole proceeds, making room for polyethylene pipe which is towed behind the mole.

For sewers of man-entry size there are some additional options. Many larger sewers are of brickwork and at the least need re-pointing. Lining the invert may be worth doing at the same time.

Where structural deterioration is more serious some form of internal lining becomes necessary. This might consist of concrete placed *in situ* against formwork designed to be moved along, the work being similar to the lining of a tunnel in new construction. Modern materials offer alternatives which minimise loss of diameter and save money though not always providing so much additional strength. These techniques involve the installation of relatively thin linings (10 to 40 mm) pre-formed from materials such as glass-reinforced cement (GRC), or plastic (GRP) or polyester resin concrete (PRC). Sections or segments of lining small enough to be man-handled are placed so as to leave a small annular space which is later filled with grout which will also penetrate the joints of the existing sewer and fill any voids which it can reach in the surrounding ground. The lining, depending on its thickness, material and form may have a permanent structural role, acting compositely with the grout and the old construction. Alternatively the lining may be intended to provide all the necessary strength or, at the other extreme, serve merely as formwork for grout and not be relied upon to provide any long term strength. Thin flexible linings need to be strutted temporarily while grouting is in progress. Some lining materials are so smooth that it is advisable to roughen the invert to provide a non-slip surface, though this is not always possible.

An ingenious process for lining pipes from 100 mm to over a metre in diameter is Nuttall's 'Insituform'. The material is terylene needle-felt impregnated with polyester resin. From one to six layers of felt, depending on sewer diameter, are pre-formed into a tube at the factory and protected inside and out with polythene sheet. The tube is delivered to the site (in a refrigerated vehicle if the temperature is high) with a length, equal to the depth of the sewer below ground, folded back over the end of the tube. The end of the folded portion is attached to an inversion collar at ground level and water, introduced to the annular space between the two layers of the tube, then gradually draws the main part of the tube through the inversion collar and forces it along the sewer by way of a vertical bend, the tube unfolding against the wall of the sewer. Once the lining is in position throughout the pipe, the resin is caused to set by circulating hot water. Installation is performed quite quickly but a great deal of preparatory work is needed to clean the sewer, remove protrusions and patch serious defects. Lateral reconnections are often made by excavation from the surface but can sometimes be opened from the inside as their positions can be seen as dimples in the lining. Remotely controlled cutters with closed circuit television have enabled re-opening to be done in pipes of quite small diameter.

Trenchless renovation of sewers, and other buried pipes, is a developing field, as is trenchless new construction. In matching the particular requirements of each job with the techniques and materials currently available some of the factors to be considered are: the extent and nature of preparatory work, whether by-pass pumping will be needed, whether extra shafts or lead-in trenches will be needed, how laterals are to be re-connected, and the effects of all these factors upon the disruption of road traffic.

The Water Research Centre's *Sewerage Rehabilitation Manual*[14] provides a systematic approach to the assessment of sewerage systems and to the planning and implementation of renovation strategies. Not least of the present problems is to decide how the funds which it is reasonable to make available can most effectively be spent, recognising that deficiencies in sewerage systems may be hydraulic or structural, or both hydraulic and structural. The Wallingford Simulation Procedure[15] (Section 2.12) has been developed to enable hydraulic checks to be carried out.

References

1 Clark, N.W.B., The causes and prevention of fractures in salt-glazed ware and other ceramic pipelines, *Public Works and Municipal Services Congress*, 1958.

2 Clark, N.W.B., The development and use of modern flexible joints for underground pipelines, *J. Inst. Pub. Hlth Engrs*, **63**, Pt 2, 108–139.

3 Ministry of Housing and Local Government, *Working Party on the Design and Construction of Underground Pipe Sewers, 2nd Report*. H.M.S.O. London, 1967.

4 Walton, J.H., *The determination of fill and surcharge loads on glazed vitrified clay pipelines*, Parts 1, 2 and 3. National Salt-glazed Pipe Manufacturers Association, London.

5 Transport and Road Research Laboratory, *A guide to design loadings for buried rigid pipes*, O.C. Young and M.P. O'Reilly, H.M.S.O. London, 1983.

6 Clark, N.W.B., Loading charts for the design of buried pipelines, *National Building Studies Special Report* No. 37. H.M.S.O. London, 1966.

7 Walton, J.H., The negative projection case in Marston's theory of external loads on buried pipelines, *J. Inst. Pub. Hlth Engrs*, **63**, Pt 1, 54–68.

8 Clark, N.W.B. and Young, O.C., Loads on underground pipes caused by vehicle wheels, *Proc. Inst. Civ. Engrs*, **21**, 91–114, Jan. 1962.

9 Water Research Centre, *Civil Engineering Specification for the Water Industry*, 2nd Edn WRC, Marlow, 1984. See also Scottish Development Dept. *Standard Specification for Water and Sewerage Schemes*, 2nd Edn, H.M.S.O., 1979.

10 Donkin, T.B., The effect of form of cross-section on the capacity and cost of trunk sewers, *J. Inst. Civ. Engrs*, 7, 261–286, Dec. 1937.

11 Humphries, C.F., The construction of sewers in tunnel and heading — the practical problems, *J. Inst. Mun. Engrs*, **93**, 333–337, Oct. 1966.

12 Kirkland, C.J., Up-to-date methods of tunnelling, *The Public Health Engineer*, 9, 42–50, Jan. 1981.

13 Snook, W.G.G., Submarine pipeline techniques for river crossings, *Civ. Eng. Pub. Wks Rev.*, **61**, 471–473, April 1966.

14 Water Research Centre, *Sewerage Rehabilitation Manual*, WRC, Marlow, 1984.

15 Department of the Environment, National Water Council, *Design and Analysis of Urban Storm Drainage, the Wallingford Procedure*, 4 vols, NWC, London, 1981.

4

Manhole chambers and storm overflows

4.1 Design and construction of manholes

A typical manhole constructed in brickwork is shown in Fig. 4.1. The base concrete completely fills the bottom of the excavation. Conventionally it is shown as extending 75 mm outside the walls of the chamber, though in fact it will often extend further. Measurement of excavation and concrete, for payment, is based on the plan dimensions shown on the drawing, the contractor allowing for any excess over these dimensions in fixing his rate. In the figure, base concrete is shown up to springing level (horizontal diameter) of the pipe. Some engineers prefer to show base concrete up to invert level only.

Manhole brickwork is always laid in English bond, courses of headers and stretchers alternating. The fair face is on the inside and 1:3 cement sand mortar is used with second-class engineering bricks. To ensure water-tightness, the brickwork must be laid to engineering rather than building standards, all joints being well flushed up with mortar, no internal voids being left. Where pipes pass through the walls, relieving arches are constructed of two rings of brickwork for smaller pipes, three rings for larger ones. The overall dimensions of brick manholes are given nominally so as to indicate that whole bricks are to be used as far as possible.

The top of the benching concrete is at pipe soffit level and is sloped towards the channel. Often the specification for benching concrete differs from that for base concrete. Thus the maximum size of aggregate might be 18 or 25 mm for the former and 40 mm for the latter. The mix would be 4:2:1 for base concrete but might be richer for the benching.

The channel is most conveniently formed by using half pipes but some engineers specify first-class engineering bricks. These are used to form both the invert and the vertical walls of the channel which are surmounted by bull-nosed bricks on edge.

The roof slab is of reinforced concrete, designed to carry traffic and earth overburden. For the typical slab a single heavy-wheel load is taken as being uniformly distributed over the clear span and the earth load added. As the slab is supported on all four sides, the usual method for 'two-way' slabs should be used in calculating bending moments. Shear is seldom critical.

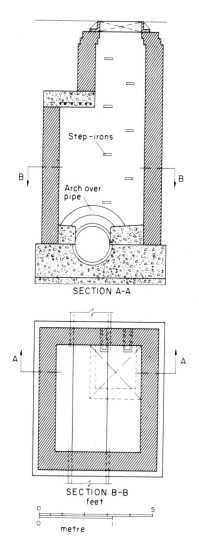

SECTION A-A

SECTION B-B
feet

metre

Fig. 4.1 Brick manhole

The square hole left in the corner for the shaft presents a difficult structural problem fundamentally but this is often overcome simply by concentrating the steel which would have crossed the hole had it not been there under the shaft walls. An alternative is to design two 1-ft wide strips under the shaft walls, crossing at right-angles as though they were beams, putting normal slab reinforcement elsewhere. The use of mesh reinforcement has many practical advantages over individual bars but bars are needed to supplement it under the shaft walls.

Some engineers use brick arches instead of concrete roof slabs. A spring-ing beam is needed to carry the shaft wall parallel to the arch axis and for the arch to abut against. This practice was more common when cement supplies were difficult.

The shaft needs to be three bricks square internally to allow easy descent, with step-irons staggered at intervals of 0.3 m apart vertically. At the top the size is reduced to 0.56 m square to accommodate the manhole cover and frame. The size is reduced by means of oversailing courses on the three sides other than that occupied by step-irons.

A wide range of types of manhole cover is available but it is not uncommon to use B.S. patterns. The most important criterion is that the cover shall be strong enough to withstand traffic loads. There are several classes of cover to suit different circumstances. Covers should not rock under traffic and triangular ones resting on three points are popular. Large square covers are made in two halves separated on a diagonal and loosely bolted together so that each half is supported at three points. A third requirement is that the cover should be of such a size and shape that it cannot be accidentally dropped through the frame. This is the main reason for the rarity of rectangular covers though they are often used for inspection chambers on drainage rather than sewerage systems.

It is not considered necessary for manhole covers to be gas-tight as sewers are adequately ventilated via the vent pipes of individual house drainage systems. Occasionally, gas-tight covers might be warranted on very long outfall lengths or where odorous trade waste is present.

Some proprietary types of cover have special advantages. One has cams operated by a special key on a recessed square nut. These secure the cover when closed. On turning the cam to release the cover, the cam provides an upward force to overcome the initial resistance. This avoids the considerable amount of hammering and prising which is sometimes needed to release an ordinary cover.

Another type consists of a steel cellular cover which is filled with concrete after laying. These are much lighter for transport and are especially suitable in footways where they look neat and provide a non-slip surface. The covers are designed to fit the frames with close tolerances and are watertight. They have to be carefully handled before installation or the fit is impaired. Special tools incorporating screws are used to release them and serve as temporarily fixed handles.

A disadvantage of the proprietary types is the necessity to use special keys. In many districts different types have been in vogue under different engineers and it is necessary to carry an armoury of devices when making inspections. Standard covers can be lifted by using a couple of picks though this should not be done too often as the key-holes become worn.

To suit different pipe diameters and depths most authorities have a range of standard manhole designs. Half a dozen different designs cover most circumstances. Some measure of standardisation of dimensions is advan-tageous so that formwork for roof slabs or arches can be used repeatedly but too few versions may lead to diseconomies in materials. Typical plan dimensions are shown in the figure.

Ideally there should be 1.85 m of headroom above the benching but this is rarely achievable where the sewer is at an economic depth. For shallow manholes or small sewers, a small inspection chamber is used as for house drains. Unfortunately most sewers are laid at depths between the maximum for this type and the minimum for reasonable headroom with a roofed chamber and shaft. The shaft itself provides unlimited headroom in these cases and, as rodding is usually carried out from the upstream end of a sewer length, the shaft is most conveniently placed in an upstream corner of the chamber. To give adequate standing space without making the chamber unduly wide, the channel is off-centre.

4.2 Precast-concrete manholes

A typical example is shown in Fig. 4.2. The chamber and shaft are formed from precast tubes with ogee joints. The chamber roof is either a precast slab with a circular hole for the shaft or, if more depth is available, a tapering section. The shaft is capped with a precast block on which the manhole frame is mounted.

With the form of construction shown, the lower chamber rings are specially ordered with pipe openings in the required positions. It is usual to prepare a schedule of these as one of the contract drawings. Surveying and setting out have to be done carefully and changes due to the presence of water mains, etc., found during excavation sometimes cause difficulties. The complications of having purpose-made chamber rings have been avoided by one authority at least by constructing the part of the chamber below the pipe crown in brickwork. Though the brickwork is circular on plan there is no need to use special bricks as they are covered by the benching concrete.

Precast-concrete manholes can be constructed with fewer man hours of skilled labour than brick manholes. Their use is much more suited to the tempo of modern construction using machine excavation for trenches and flexibly jointed pipes. Brick manholes are becoming obsolete except for special purposes, such as overflow chambers.

4.3 Deep manholes

At depths greater than about 3 m, 225 mm brickwork is generally considered to be inadequate and the thickness is increased to $1\frac{1}{2}$ bricks. At about 4.5 m depth it is increased to two bricks. Different engineers specify different depths for increases in thickness. Concrete tubular manholes are surrounded with 150 mm of *in situ* concrete at depths greater than about 1.8 m.

Steel ladders are used instead of step-irons in deep manholes and rest platforms are provided at intervals. In very deep manholes it is advisable to

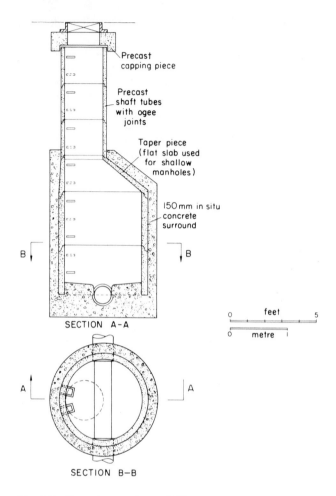

Fig. 4.2 Precast-concrete manhole

stagger successive lengths of ladder so that there is no possibility of a man falling the full depth, though safety harness should be worn. Where intermediate platforms are provided they should be arranged so that it is possible to raise a bucket or small skip directly from the bottom. This will involve having two small holes in each platform, one for man access and one for skips. The former will be staggered on successive platforms and the latter will be vertically below each other and should have guard-chains or rails. Shafts in such cases need to be quite large and may need multiple covers. Since deep sewers are usually constructed in heading, the manholes are built in the working shafts and are often constructed of bolted segments of precast concrete, circular on plan, fixed as the excavation proceeds.

4.4 Manholes on large sewers

An example is given in Fig. 4.3. If the sewer is deep enough, the floor of the chamber can be at a level near that of the top of the sewer, but if it is fairly shallow 1.85 m of headroom cannot be provided without accepting a wet chamber when the flow is high. This is not unduly serious as the primary purpose of the manhole is to give access for de-silting, inspection and repair during periods of dry-weather flow.

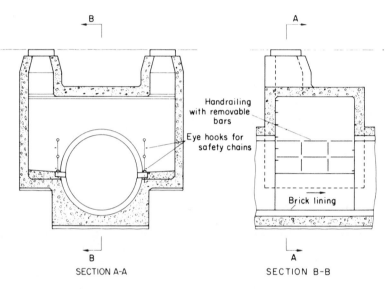

SECTION A-A   SECTION B-B

Fig. 4.3 Reinforced concrete manhole on a large-sewer

Safety rails at the edge of the platform are essential. At conventional manholes on sewers of 0.75 m diameter and above provision should be made for fixing safety chains across the sewer at the downstream end of the manhole.

4.5 Drop manholes

The most common application of drop manholes was in places where, in order to limit velocities, it was necessary to lay sewers at much flatter gradients than those of the land, Fig. 4.4. As sewerage engineers become more confident in employing high velocities and steep gradients, drop manholes may become less common though there will still be cases where a local very steep slope is most easily dealt with by a vertical sewer drop. Side connections to deep sewers often require vertical drops.

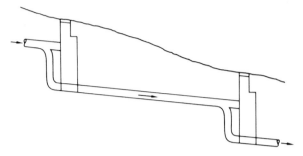

Fig. 4.4   Use of back-drops to avoid steeply sloping pipes

Figure 4.5 shows a typical 'side-drop' manhole on a small sewer. The bottom of the drop is a cast-iron duck-foot bend. The duck-foot makes building easier but is not absolutely necessary. In the figure, a 'tumbling-bay' junction is used for the top of the drop. This differs from a standard junction in that it has spigots in place of the latter's sockets and vice versa. An alternative is to use a standard 45° junction with a 45° bend together with a double socket to connect the junction to the upstream pipe run. Even if a tumbling-bay is used a double socket will be needed somewhere on the drop to connect concrete to cast-iron. An iron spigot can be jointed to a concrete or glazed-ware socket but iron sockets are too small to take concrete or glazed-ware spigots. There is plenty of scope for considering alternative arrangements.

Some designers prefer to accommodate the whole of the drop within the chamber, constructing it entirely in cast-iron except for the tumbling-bay which is constructed in concrete and left open for access. A compromise is to provide an upper tumbling-bay chamber but to have the drop-pipe itself outside the lower part of the chamber. Deep drops are sometimes done in stages.

The vortex drop (p. 149) is becoming popular for medium-sized sewers, perhaps because its behaviour is predictable. It is evidently a good way of dissipating some of the energy of the falling water, though it is possible that ordinary drop-pipes behave in a similar manner.

When a large sewer has to be dropped, a ramp is used. An alternative is a cascade of steps. The advantage claimed for cascades is that the energy is dissipated a little at a time but this is probably true only at low flows when the problem is not serious. Steps are certainly subject to more wear and if they are constructed of brickwork very frequent inspection is desirable. Once a brick is loosened, deterioration is rapid.

With a ramp (or cascade except at low flows), a hydraulic jump (see p. 143) will occur on the ramp or at the tail over a wide range of flow in most cases, dissipating a good deal of energy. If the sewer is large, investigations by calculation or scale model tests may be warranted.

The energy to be dissipated at the foot of the ramp depends on the gradient and is almost independent of the overall drop if the ramp is long enough for a

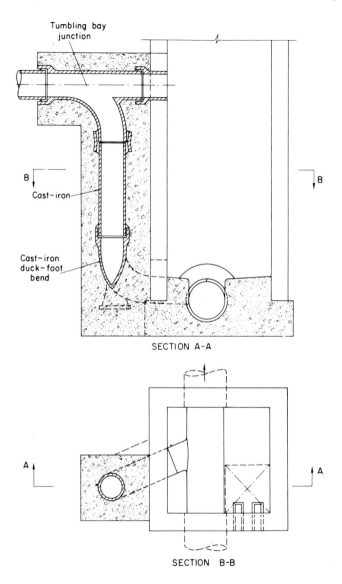

Tumbling bay
junction

B

Cast-iron

Cast-iron
duck-foot
bend

SECTION A-A

A

SECTION B-B

Fig. 4.5 Drop-junction manhole

supercritical normal depth to be approximately reached (see Chapter 7).
Energy due to the fall in elevation is being continuously dissipated by friction
along the ramp. The energy remaining at the foot is mainly the kinetic
energy ($V^2/2\,g$) due to the high velocity caused by the steep slope.

An interesting alternative to a straight ramp is a helical one which can be accommodated in a cylindrical chamber. This occupies much less length but there is a vortex action which tends to pile up the water against the walls. A drop of this kind was suggested by K.O. James for a scheme at Ashton-under-Lyne. In that instance model tests showed that the idea was scarcely feasible owing to extremely high velocity in the existing sewer upstream, but there are undoubtedly circumstances where it would be a neat solution. A helical step cascade for small sewers is illustrated in the paper describing the West Middlesex Scheme[1] and was found to work quite well.

Sewers of intermediate size (1 m diameter) are rather large for vertical back-drops and too small for a specially constructed ramp to be warranted. Inclined drops formed from pipes, with manhole chambers at both ends, are used for these. As mentioned above, the vortex drop is an alternative. The ramp is suitable for small drops, the vortex for larger ones but there is a considerable overlap when either type is feasible and it is in this range that a helical cascade is also worth considering.

4.6 Storm overflows

The need for storm overflows arises because the peak flow in combined sewers during severe storms is so large that the conveyance and treatment of the full discharge would be impracticable. The earliest forms of overflow were intended only to restrict the flow to downstream sewers, surplus flow being spilled to a watercourse. Later developments have been directed toward attempting to retain foul matter in the through-flow especially during the early part of a storm when the scouring action of the flow may result in a high degree of pollution.

The several types of overflow chamber are described below in their logical order of development. Consideration of design factors will be discussed later.

4.7 Early forms of overflow

The principle of the leaping weir is shown in Fig. 4.6. Low flows simply fall to the floor of the chamber and pass to the through-flow pipe. At high flows the trajectory is longer and a large proportion of the flow is caught by an adjustable trough and diverted to overflow pipes which lead to the watercourse. If the trough is set to give a chosen maximum through-flow for moderate storms, the through-flow will be less for severe storms. If the setting is for severe storms, through-flow will be too large in moderate storms. No attempt is made to prevent gross solids from passing to the watercourse though it is possible that heavy inorganic particles carried as bed-load would remain in the through-flow.

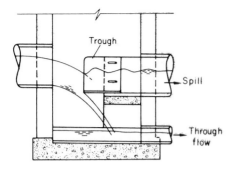

Fig. 4.6 Leaping-weir overflow

The low side-weir overflow is shown in Fig. 4.7. The weir-sill level is set at about normal depth for the chosen rate of through-flow in the downstream pipe. These overflows were customarily designed with the aid of a formula proposed by Coleman and Dempster–Smith.[2] The great majority of practical cases lie beyond the range of the originators' experiments, with the results that first spill usually occurs at a discharge less than intended while the through-flow is much greater than intended at the peak of the storm. In spite of the dip-plates there is a high risk of discharging polluting matter to the watercourse. Though more exact analysis of side-weir flow is now possible,[3] later forms of overflow chamber are considered to be preferable.

Many existing storm overflows are in need of replacement or even elimination. Apart from leaping weirs and low side-weirs there are many overflow chambers which consist of little more than a relief pipe built into the wall of an existing manhole above benching level.

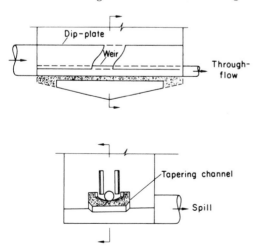

Fig. 4.7 Low side-weir overflow

4.8 Stilling-pond overflows

The original form of the stilling-pond overflow[4] is shown in Fig. 4.8. Low flows pass straight through the chamber. The diameter and length of the throttle or module pipe are such that the chosen rate of through-flow for first spill is reached when the pond is full to weir level. The weir is set at a level not far below the soffit of the incoming sewer so that both the pond and the sewer have to fill up before spill commences. During the filling period, some storm flow passes to the downstream sewers. At the end of the storm, the contents of the storage volume drain away. In each storm, the volume of flow discharged to the watercourse is less than the volume of stormwater entering the chamber. The difference is equal to the combined storage volume of the pond and the upstream sewer. Storms of smaller volume than the storage capacity will not cause spill. Thus both the occasions and the volume of spill are reduced.

At high rates of spill the through-flow will be greater than at first spill by a factor of

$$\left(1 + \frac{h_w}{h_0} \right)^{\frac{1}{2}}$$

This factor can be limited by making h_0 large or by using a long weir to make h_w small. Site limitations often restrict h_0 and, since the chamber floor must have adequate side slopes to avoid deposition, this will also restrict the width of the chamber. An adequate weir length has sometimes been provided by returning the weir along the sides of the chamber.

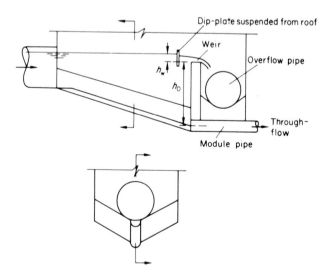

Fig. 4.8   Stilling-pond overflow

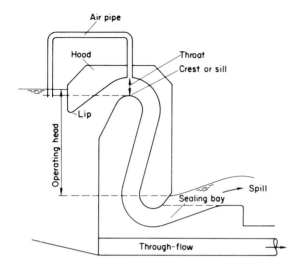

Fig. 4.9 Intermittent siphon as an alternative to an end-weir in a stilling-pond overflow

The use of siphons saves space while at the same time restricting h_w. Figure 4.9 shows the form of siphon typically used in early stilling-pond overflows. The hood seals the inlet when water level approaches the sill and acts as a barrier for floating solids. Water spilling over the sill entrains air in striking the opposite wall or in plunging into the pool retained in the sealing bay. This exhausts air from the space under the hood and the siphon should prime and run full before the water level has risen very far above the sill. Once primed the siphon discharges a fixed quantity:

$$Q = CA\sqrt{(2gh)} \tag{4.1}$$

in which h = difference between the water level at inlet and the level at which flow leaves the solid boundary at the tail,

A = cross-sectional area of the siphon (if the area varies, the throat area is taken), and

C = a coefficient covering the conversion of potential to kinetic energy and the losses of energy. The value of C usually lies between 0.5 and 0.8.

When the approaching flow is less than the discharge, the upstream water level will be drawn down. To prevent siphonic action continuing until the lip is exposed, when there would be a risk of drawing in floating material, air pipes are provided. These extend to a level slightly below the sill. When they are exposed air passes into the hood and siphonic action is stopped. The siphon 'hunts', that is it operates intermittently, at all flows below its design capacity. Where several siphons are used they are sometimes placed with their sills at different levels so that they will not all hunt at once.

An orifice may be used instead of a module pipe but a pipe is usually preferable as it can be larger diameter for the same head and discharge. Oakley[5] recommends that orifices or pipes should not be less than 225 mm in diameter to avoid the danger of blockage. The Venturi flume is an alternative module but is unsuitable for small overflows and aggravates the problem of minimising excess through-flow since the factor becomes

$$\left(1 + \frac{h_w}{h_0}\right)^{\frac{3}{2}}$$

This form of control is very useful at treatment works (Section 9.12) where mechanical or electrical gear can be used and may be worth considering for large overflows.

The vortex drop is another possible form of control. The excess through-flow factor is approximately $1 + h_w/h_0$. If enough fall is available for a vortex drop it is usually better to use a deeper stilling-pond with orifice or pipe control.

Sharpe and Kirkbride[6] carried out many experiments to find proportions for stilling-pond chambers which would restrict the passage of gross floating and suspended solids over the weir. Their recommendations are shown in Fig. 4.10. In contrast to earlier designs, the incoming sewer is surcharged when spill occurs. To restrict the length of surcharging pipe, a drop in invert

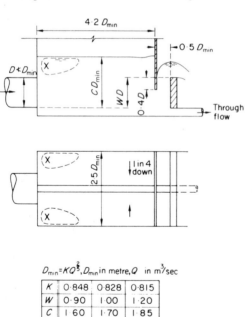

$D_{min} = KQ^{\frac{2}{5}}, D_{min}$ in metre, Q in m^3/sec

K	0·848	0·828	0·815
W	0·90	1·00	1·20
C	1·60	1·70	1·85

Fig. 4.10 Stilling-pond proportions recommended by Sharpe and Kirkbride. Solids accumulate in zones marked X

level can be provided upstream. The drop would be of the same order as that within a chamber of the earlier form.

In a chamber of the recommended form, eddying but relatively quiescent zones occur as shown in the figure. The gross solid matter rises behind the dip-plates and moves towards these zones where a good deal of it is retained until spill ceases.

A minimum diameter, D_{min}, is specified for the incoming pipe because it was found the efficiency of retention fell abruptly when the Boussinesq number, $V/\sqrt{(2gm)}$, for the flow in the pipe fell below 0.85. A pipe of larger diameter than D_{min} may be used but in that case chamber length and width, and water level are still based on D_{min}. In contrast the dimensions relating to the dip-plate are based on the actual diameter of the pipe.

It appears that the through-flow at peak spill will be about 30% in excess of that at first spill with these proportions and that the combination of dip-plate and weir wall will act as a slot-shaped orifice. The relationship of plate to weir was adopted after some preliminary experiments and unaltered in the main series.

Some doubts have been expressed about the significance of the Boussinesq number and there are other features of the problem which provide scope for further work. Sharpe and Kirkbride found that quite small changes could affect performance. If site limitations prevent their recommendations from being adopted in full a risk of less satisfactory performance arises and consideration should be given to carrying out model tests.

Markland and Frederick[7] experimented with model stilling-ponds which had siphons instead of weirs. They used a wide range of suspended objects, and examined the theoretical implications of their results. Their design is shown in Fig. 4.11.

The siphon is of such a form that it will discharge steadily over a wide range of flow either as a sub-atmospheric weir or with a mixture of air bubbles and water. The hood is shaped to encourage the entrance of air and does not need to act as a scum barrier since a high proportion of gross solids is retained behind the deep lateral scum-board. The incoming pipe is again submerged when spill occurs. The dry-weather flow channel may be partially covered so that heavy material is retained in the through-flow.

Recommended minimum proportions are $L/D = 7$, $H/D = 2$, $B/D = 2.5$. Larger ratios will give better performance, and a method is offered for estimating the degree of improvement. The scum-board is placed to make $l/L = 0.75$ and $h/H = 0.5$ or 1.25 if H/D is larger than 2.5. The siphon is placed as far downstream as possible. One or more siphons may be installed symmetrically in the downstream end wall.

4.9 High side-weir overflows

This type of overflow is unlikely to be adopted except where it is quite impossible to arrange for a drop in sewer level through the chamber. Nevertheless

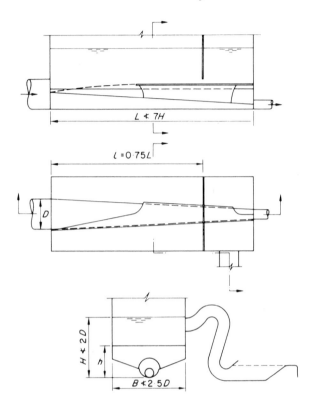

Fig. 4.11   Stilling-pond with siphon overflow developed by Frederick and Markland

it appears to function fairly well and particular models have been reasonably successful in retaining a proportion of the pollutants.[8,9]

In principle, the high side-weir overflow, Fig. 4.12, is similar to the stilling-pond type but because there is little or no drop through the chamber, and the flow has to be self-cleansing, the 'pond' can be little wider than the incoming sewer. There is insufficient width for an end weir so side weirs are used instead. The through-flow is controlled by an orifice, module pipe or other device as in a stilling-pond and full use is made of upstream storage capacity before spill commences. Siphons could be used in place of the weir but might cause more solid matter to be drawn from the through-flow. It is possible that a deep lateral scum-board upstream of the weirs or siphons would be of benefit but experiments would be needed to confirm this.

A form of high side-weir chamber providing for storage of floating solids has been developed by laboratory model testing but field trials of an adaptable prototype would be advisable before adopting it widely.[10]

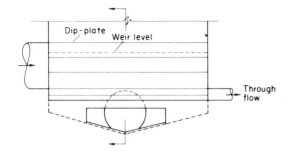

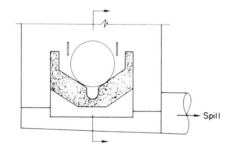

Fig. 4.12 High side-weir overflow

4.10 Vortex overflows

These should not be confused with the vortex drop which, as noted elsewhere, can be used as a downstream control for an overflow chamber. The vortex overflow was developed at Bristol where prototypes have been constructed.[11] Models have been compared with other types in the laboratory and in the field.

The principle of operation can be followed from Fig. 4.13. Flow enters the basin tangentially and circulates round the weir wall. In experiments with small-scale chambers it was found that solid matter tended to accumulate round the base of the weir wall. The entrance of the through-flow pipe is therefore adjacent to this wall. A channel is provided to guide low flows to the through-flow pipe. A guide wall near the incoming sewer is needed, as a fully circular chamber would give rise to radial waves at certain discharge rates (compare Section 7.8 — perhaps a volute-shaped chamber would be advantageous). A scum-board is needed round the weir sill. Floating matter is retained more successfully when the scumplate is not too close to the weir.

If sufficient fall is available between the incoming and outgoing sewers, the chamber may be deeper. For a given amount of fall, there would usually be more storage capacity in a vortex overflow than in a stilling-pond.

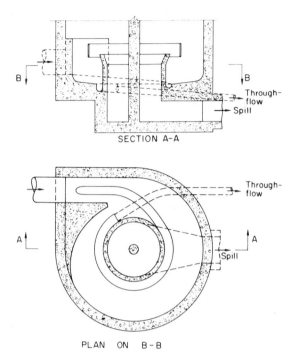

SECTION A-A

PLAN   ON   B - B

Fig. 4.13   Vortex overflow

4.11 Storage overflows

The several types of stilling-pond overflow, the high side-weir and vortex types all employ the storage capacity of upstream sewers to reduce amounts, frequencies and volumes of spill. Development and investigation of these types since Braine[4] drew attention to the value of storage have been directed mainly towards attempting to promote selective operation by concentrating polluting matter in the through-flow. Experimental work has usually involved introducing fairly large particles of suitably contrived rising or settling velocity into the flow. However, field observations in Birmingham[12] showed that a great deal of polluting matter was in finely divided or dissolved form and was therefore unlikely to be retained in the through-flow selectively. It was found that a large proportion of the additional polluting matter during a storm occurred during the time up to the peak, as approximated by the time of concentration. It was recommended that storage for a corresponding volume of flow should be deliberately provided. Flow subsequent to the peak would spill to the watercourse once the storage was full. The storage was to be provided off-line, the sewer having two overflow weirs, a lower one spilling to the retention tank and an upper one spilling to the

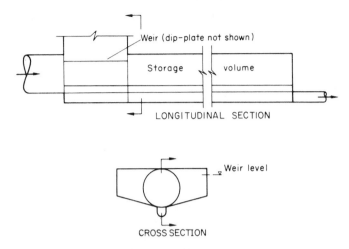

Fig. **4.14** Storage overflow. With so flat a floor slope excessive silting might occur. If possible the length would be increased and the width reduced, or the through-flow pipe would be lowered

watercourse. Off-line storage was practicable at sites where existing overflows had steep drops downstream as the tanks could then drain by gravity.

Elsewhere, on-line storage has been more common, usually being provided downstream of the weir, Fig. 4.14, so that most polluted flow has passed the weir before spill begins.

Methods similar to those for balancing ponds (Sections 2.13 and 2.14) can be used to determine storage capacity and other features at the design stage, the objective being to retain the volume of flow accumulating over a period about equal to the time to the peak of the hydrograph or even to the duration of the storm used to derive the hydrograph. Figure 4.15 shows how the inflow, through-flow and spill vary. There are minor differences in the through-flow curve according to whether the storage is on-line or off-line.

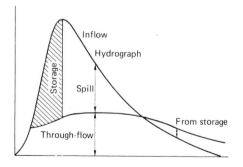

Fig. **4.15** Operation of an off-line storage overflow. Through-flow has been exaggerated relative to peak inflow

To the extent that the demand for higher standards of stream protection now warrants the additional expense of providing storage overflows, the earlier types may become obsolete except in locations where the spill is to a large river or estuary. Further work at Birmingham[13] established standards for settings and storage provision which were related to frequency of operation and the amenity value and size of the receiving stream. It was suggested that for overflows discharging to amenity watercourses storage should be based on twice per year storms and for other watercourses on ten times per year storms.

4.12 Off-sewer storm tanks

These tanks are designed to receive all flow spilled at the overflow chamber and act as sedimentation tanks, so that sewage passing through them will be clarified before discharge. The contents of the tanks, including the sludge from the floors, are pumped into the downstream sewer after the storm. Many storms will be insufficient to completely fill the tanks.

A system on this principle has been installed at Coventry[14] where the receiving watercourses are small. Automatic equipment is needed to avoid continuous attendance.

Though the principle of using off-sewer tanks is similar to that of providing storm tanks at sewage works it should be noted that, while the latter have usually to deal with flows up to only 3 × DWF, much greater flows have to be passed through off-sewer tanks. Storm tanks at sewage works are normally given about 2 hours retention capacity at maximum rate of flow. It will rarely be feasible to provide for so long a detention period in off-sewer tanks.

4.13 Storm overflow settings

In a combined system, since storm water and polluted water are mixed together, the whole flow becomes polluted. Storm overflow chambers are incapable of retaining more than grit and coarse solids so some polluting matter is inevitably discharged to the stream. The through-flow which the chamber can pass without spill is commonly known as the setting.

For many years it was customary to set overflows to come into operation at 6 × DWF. This criterion seems to have been adopted indirectly from a recommendation of the Royal Commission that treatment works should be designed to fully treat flows up to 3 × DWF and to provide storm tank treatment for flows up to a further 3 × DWF. Thus, if works were provided for flows up to only 6 × DWF any surplus above that necessarily had to be overflowed from the sewers without treatment.

The justification for the practice of overflowing sewage without treatment during storms has always been that in high flows polluting matter is considerably diluted. Thus if the mean dry weather concentration of a pollutant (measured for instance as BOD, suspended solids, etc.) is b and the mean dry weather flow is q, the mean concentration during a storm flow of Q is $b\,q/Q$. The greater the storm flow the less the concentration of the sewer flow and hence of the spill. The highest concentration which can occur in the spill, neglecting hour-to-hour changes in pollutant discharge to the sewers, is $b\,q/Q_t$ when Q_t is the over-flow setting. An $n \times \mathrm{DWF}$ rule makes Q_t equal to nP $(w + i)$ or to $P(nw + i)$ in which P is the population, w is the daily dry-weather polluted water flow per head and i is the daily infiltration flow per head. The two versions have been given because of different interpretations of DWF (see page 10) when used with a multiplier. The limiting concentration would then be

$$\frac{bq}{Pn\,(w + i)} \text{ or } \frac{bq}{P\,(nw + i)}$$

In either case if an invariable value of n is used, the occurrence of different values of w (or indeed of i) will result in different standards in different areas. Consequently, it may be better to specify a value for $n(w + i)$ or of $(nw + i)$ rather than simply for n. During the nineteen-forties such a practice tended to replace the use of $6 \times \mathrm{DWF}$ as the setting criterion, the allowance being 240 gal/d that is 1090 l/d per head of population. Whether this allowance was considered to be $n(w + i)$ or $(nw + i)$ is immaterial. Incidentally, with a typical daily BOD contribution of 0.055 kg/head, the limiting concentration is $0.055 \times 10^6/1090$ or approximately 50 mg/l, which appears to be not unduly high in comparison with the 20 mg/l frequently adopted as a treatment works effluent standard.

Industrial effluents can be taken into account by estimating the additional population which would produce a daily mass of pollutant equal to that of the industrial effluent.

There is a serious defect in using criteria based on simple dilution because the concentration of pollutants, especially of suspended solids, is very much higher in the early part of a storm.

In 1955 the Minister of Housing and Local Government appointed a Technical Committee on Storm Overflows and the Disposal of Storm Sewage. Their final report was issued in 1970 and contains a very thorough review of the problems associated with storm discharges from combined systems, together with design recommendations.[15,16,17,18]

From field studies which they commissioned the committee was able to evolve a number of possible criteria for overflow settings. The setting could be fixed to limit the duration, the frequency, or the volume of spill from an overflow. These factors are of course inter-related and are basically hydrogical. A further criterion also pursued by the committee concerned the mass of pollutants spilled per annum, which could be compared with the mass discharged by the treatment works per annum.

Settings based on any of these criteria achieve much the same effect. Thus a high setting reduces the frequency of spill (since more light storms will be passed without spill) and inevitably reduces the duration of spill as well as the amount of both water and pollutants spilled. However, an overflow which employs the available storage capacity of the sewers (such as a stilling-pond type) or a scheme involving extra storage possibly downstream of the weir (Fig. 4.14) also reduces the frequency of spill so storage provisions in addition to setting enters into considerations of criteria. This aspect is not however explored in the committee's report.

The majority of the committee finally agreed that a fairly simple approach to the design criterion should be recommended and the following formula was given for the maximum rate of flow to the treatment works from combined and/or partially separate sewerage systems:

$$\text{DWF} + 1360\,P + 2E \quad \text{litre/day}$$

DWF is the mean rate of flow in the sewers in dry weather and is therefore the sum of the mean rates of flow of domestic polluted water, industrial effluent (E) and infiltration (I).

P is the population of the area drained to the overflow site on the combined and partially separate system.
E is the volume of industrial effluent discharged in 24 hours.

The formula may be expressed alternatively as

$$P(w + 1360) + 3E + I$$

Where part of the area drained to an overflow was on the completely separate system 3 × DWF from this area would be added (P in the above formula applies only to combined and partially separate areas).

It was envisaged that in special cases such as discharge to very small or very large rivers, the 1360 l/d per head figure might be raised or lowered. Abnormal circumstances of industrial effluent discharge might justify modifying the $2E$ term.

It will be noted that the committee's formula, by having a prominent term involving the population, avoids the original defect of the 6 × DWF rate (that it leads to standards varying with water consumption) but nevertheless allows water consumption (and infiltration) to contribute to the final result via the DWF.

The formula does not, however, involve any direct reference to concentration of pollutants nor to the polluting effects of discharges on receiving water courses. It is arguable that in especially sensitive cases detailed investigations should be carried out and consideration given to frequencies, volumes and strengths of discharges; much helpful information will be found in the Technical Committee's final report. The influence of storage also needs to be borne in mind.

Though the lowest overflow on the system must be set to restrict the flow to the treatment works to that given by the Technical Committee formula (modified if necessary) there is much to be said for adopting higher settings further upstream. The economic consequences of adopting higher settings are often not serious when the sewer receives further storm water downstream of an overflow, as the through-flow from the overflow chamber can be insignificant in comparison with the added storm flow. Model studies suggest that high ratios of through-flow to spill flow are beneficial in retaining suspended matter.

Finally, it must be emphasised that the hydraulic capacity of the receiving stream must always be checked. If it is inadequate to take the spill, an overflow must not be installed at that point. Obvious though this may be sewerage authorities in the past have sometimes saved money on sewerage schemes by providing overflows and then had to spend money on alleviating consequent flooding of water-courses particularly at culverts.

4.14 Hydraulics of storm-overflow components

This section should be read in conjunction with Chapters 6 and 7.

(1) Weirs

The weir of a stilling-pond overflow of the form shown in Fig. 4.8 can be designed as a broad or round-crested weir according to the shape chosen. The dip-plates will create a small additional afflux. Frequently this is allowed for by making the weir about 10% longer than it would need to be without dip-plates.

If the space between the dip-plate and the weir is small relative to the head above the weir, the space acts as a long narrow slot or orifice from which a super-critical flow issues. The head for a given discharge is then greater than that calculated from a weir formula. There appears to be no simple way of predicting the discharge at which the transition occurs, nor of calculating the head for the upper mode of flow. This problem rarely arises as the maximum head on the weir is seldom more than 200 mm in overflows of the form shown in Fig. 4.8.

The Sharpe and Kirkbride overflow appears to operate deliberately above the true weir-flow range, but here the weir length does not have to be determined as it is the same as the width of the chamber.

Side-weir flow is dealt with in papers previously referred to[3] and in Chapter 7. Though low side-weir overflows would have to be designed by taking account of the varying depth along the weir, most high side-weir overflows have a channel of sufficiently generous proportions for ordinary weir formulae to give an acceptable approximation to the head for a given discharge. In cases of doubt the flow can be analysed more rigorously as a check after the weir length has been determined by assuming ordinary weir flow.

The circular weir in a vortex overflow has some similarity to a vortex drop (see Section 7.8). At low discharges lip-weir flow will occur, that is the weir will behave as a straight weir of the same length. At higher discharge it seems likely that the head will be greater than that for true weir flow due to the circulation in the vortex. Presumably the vortex-drop theory could be adapted to give at least a first approximation to the head for a given discharge in these circumstances. Experimental verification would be needed unless there were a safety margin.

(2) Siphons

The required throat area for a siphon is found by using equation (4.1). The coefficient has to be guessed, and the operating head can be only roughly defined. With a siphon the upstream water level will rise sharply if the design discharge is exceeded so it may be prudent to ensure that overtopping of the siphon can occur without disastrous results.

Many difficulties arise in scale-model testing of siphons. Discharge coefficients can be obtained reliably only by testing a prototype, though extrapolations may be attempted from the results of tests at two or more scales. Since priming involves aerated flow, surface tension and atmospheric pressure enter the problem. Intermittent operation of a siphon depends also on the surface area of the head pond which may include part of the upstream sewer. All these factors render planning and interpretation of scale-model tests so difficult that safety margins may still have to be allowed.

(3) Module pipe or orifice

The module pipe is designed simply as a pipe with a reservoir upstream and a free outlet, Fig. 4.16. Conversion of potential to kinetic energy at the inlet should not be forgotten. Pipe friction and entry loss must be guessed. From the point of view of overflow operation, it is best to overestimate the losses but when considering downstream flow it may be prudent to underestimate the losses. The pipe will need a small gradient so that velocity is adequate in dry weather.

In a situation like that shown in Fig. 4.16, the pipe does not necessarily flow full right to the open end. It is possible for the hydraulic gradient to intersect the top of the pipe upstream of the end and for one of the uppermost profiles shown in Fig. 7.5 to occur. These cases can be identified and

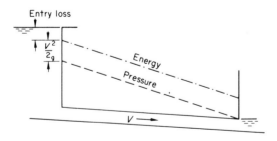

Fig. 4.16   Head conditions for a module pipe

analysed only by trial profile calculations which in practical terms are not likely to be of great accuracy. Where the gradient of the module pipe is very small as is usually the case, the flow is not likely to be seriously underestimated by assuming the pipes to flow full to the end.

The sewer downstream of the pipe will usually be of larger diameter. If head is to be conserved, a level-invert transition may be used but it will then be necessary to check whether the pipe outlet is likely to be drowned. This will depend on whether or not the specific force of the flow from the module is less than that in the free-surface sewer, which may usually be assumed to flow at normal depth. If the module outlet is drowned, the tailwater level will need to be taken into account.

Ackers *et al.*[8] noted that a module pipe from a model vortex overflow carried more flow than intended and attributed this to the circulation in the chamber.

If an orifice is used instead of a pipe for a stilling-pond or storage overflow, the discharge coefficient must be guessed. A value of about 0.65 is possibly reasonable.

Since losses can only be estimated, whether a pipe or an orifice is used, it may be considered preferable to install a throttle plate or penstock so that adjustments can be made. This will be feasible only where discharge measurements can be made at the commissioning stage. If the storage is large and there is enough water to fill it, the rate of fall method can be used. With a small storage, the fall would be too rapid.

Where an adjustable throttle is provided it should be designed so that complete closure is impossible once the overflow is put into normal operation. It may be advisable to make the position permanent and unalterable once a suitable setting has been found.

References

1 Watson, D.M., West Middlesex main drainage, *J. Inst. Civ. Engrs*, **5**, 463–617, April 1937.
2 Coleman, G.S. and Dempster Smith, The discharging capacity of side weirs, *Selected Engineering Paper* No. 6, Inst. Civ. Engrs, 1923.
3 Four papers on side weirs, *Proc. Inst. Civ. Engrs*, **6**, 250–343, Feb. 1957:
   Ackers, P., A theoretical consideration of side weirs as storm-water overflows.
   Allen, J.W., The discharge of water over side weirs in circular pipes.
   Collinge, V.K., The discharge capacity of side weirs.
   Frazer, W., The behaviour of side weirs in prismatic rectangular channels.
4 Braine, C.D.C., Draw-down and other factors relating to the design of storm-water overflows on sewers, *J. Inst. Civ. Engrs*, **28**, 136–163, April 1947.
5 Oakley, H.R., Practical design of storm sewage overflows, *Symposium on Storm Sewage Overflows*, 115–122, Inst. Civ. Engrs, London, 1967.
6 Sharpe, D.E. and Kirkbride, T.W., Storm-water overflows: the operation and design of a stilling pond, *Proc. Inst. Civ. Engrs*, **13**, 445–466, Aug. 1959, *Disc,* **16**, 283–286, July 1960.
7 Frederick, M.R. and Markland, E., The performance of stilling ponds in

handling solids, *Symposium on Storm Overflows*, 51–61, Inst. Civ. Engrs, London, 1967.

8 Ackers, P., Harrison, A.J.M. and Brewer, A.J., Laboratory studies of storm overflows with unsteady flow, *Symposium on Storm Overflows*, 37–49, Inst. Civ. Engrs, London, 1967.

9 Ackers, P., Brewer, A.J., Birkbeck, A.E. and Gameson, A.L.H., Storm overflow performance studies using crude sewage, *Symposium on Storm Sewage Overflows*, 63–77, Inst. Civ. Engrs, London, 1967.

10 White, J.B. and Hutchinson, R.E., A storm overflow chamber without drop in invert level, *J. Inst. Mun. Engrs*, **100**, 279–284.

11 Smisson, B., Design, construction and performance of vortex overflows, *Symposium on Storm Sewage Overflows*, Inst. Civ. Engrs, London, 1967.

12 Hedley, G. and King, M.V., Suggested correlation between storm sewage characteristics and storm overflow performance, *Proc. Inst. Civ. Engrs*, **48**, 399–411, March, 1971.

13 King, M.V., Planning and construction of Birmingham's main outfall sewer, *The Chartered Municipal Engineer*, **103(6)**, 87–95 (June 1976).

14 Steele, B.D., The treatment of storm sewage, *Symposium on Storm Sewage Overflows*, 23–27, Inst. Civ. Engrs, London, 1967.

15 Ministry of Housing and Local Government, *Technical Committee on storm overflows and the disposal of storm sewage: Interim Report*, H.M.S.O., 1963.

16 Davidson, R.N. and Gameson, A.L.H., Field studies on the flow and composition of storm sewage, *Symposium on Storm Sewage Overflows*, 1–11, Inst. Civ. Engrs, London, 1967.

17 Lester, W.F., Effect of storm overflows on river quality, *Symposium on Storm Sewage Overflows*, 13–21, Inst. Civ. Engrs, London, 1967.

18 Ministry of Housing and Local Government, *Report of Technical Committee on storm overflows and the treatment of storm sewage*, H.M.S.O., 1970.

5

Pumping systems and inverted siphons

5.1 Introduction

In the majority of cases centrifugal pumps are used for pumping sewage. The presence of solid matter has a major influence on design, and unless a screening and grit removal plant can be conveniently installed upstream of the pumping system, special pumps have to be used. These have impellers of simple shape, Fig. 5.1, capable of passing almost any object entering the suction connection, and have lower mechanical efficiencies than conventional centrifugal pumps.

The necessity of avoiding blockages means that pumps of suction diameter less than 100 mm cannot be used, and for flows less than about 0.01 m^3/s, pneumatic ejectors have been used instead of centrifugal pumps.

Often only the foul sewage or the through-flow from storm overflows has to be pumped, storm water being discharged to a watercourse by gravity. The duty is then such that the specific speed is within the range for a centrifugal

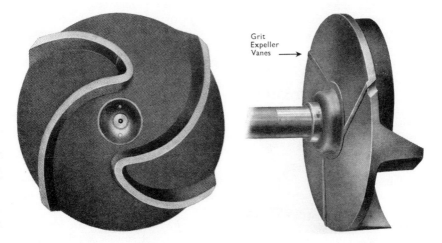

Grit
Expeller
Vanes ⟶

Fig. 5.1   Impeller for centrifugal sewage pump (*Adams-Hydraulics Ltd, York*)

pump. Where the total head exceeds about 30 m, pumps have to be coupled in series.

Where storm water has to be lifted to a watercourse and the lift is low in relation to discharge, axial-flow pumps are needed. Where the system is separate, the storm-water pumping station is similar to a land-drainage pumping station and has few features peculiar to sewerage practice. If the system is combined screens or comminuters and grit-removal tanks may be almost essential if axial-flow pumps are needed and two sets of pumps may be provided: small, perhaps centrifugal, pumps for dry-weather flow and axial-flow pumps for storm use.

Though there is occasionally a case for engine-driven pumps, electrical drive is normally considered to be more economical and reliable. If the station is large and failure would have serious consequences, two independent external electricity supplies are provided. If a second external supply cannot be obtained, a diesel generator is installed as standby. This is usually preferable to a separate diesel-driven pump.

The variation of sewage flow introduces several design problems. Since power is normally taken from the a.c. public supply and variable-speed a.c. motors are expensive, variable rate pumping is not usually economic. Instead, sewage is pumped intermittently using constant discharge pumps, a storage well being provided for balancing purposes. Float gear or electrodes start and stop pumps as the well fills and empties.

It is often desirable to install several pumps in parallel, one pump being used for low flows and additional pumps coming into service as the flow increases. A single pump large enough to deal with the maximum flow would involve long periods of storage at low flows and would create problems at the treatment works.

As the pumps run intermittently, priming difficulties cannot be overlooked and are commonly solved by placing the pumps below the lowest water level in a separate dry-well. To avoid trouble with the motors should the dry-well be flooded, vertical spindle pumps are used with motors at ground level. Direct coupled units with the motor immediately above the pump are available for shallow wells or where the risk of flooding is negligible.

The provision of a dry-well can be avoided by using a pump submerged in the sump, but maintenance is more difficult and provision has to be made for raising the complete unit. Pumps with motors attached in submersible, sealed packages are commonly used for small flows. The whole pumping station can be below ground in the form of a cylindrical chamber which serves as the well.

Pumps can be installed above water level if priming gear can be countenanced. For automatic stations, sewerage engineers rarely care to rely on exhausters but an interceptor chamber on the suction line which retains sufficient water for priming has sometimes been found acceptable.

The separate dry-well, accommodating vertical spindle pumps below sump level, remains the popular choice and the details below relate primarily to this layout, though many of the principles to be discussed apply equally to the alternatives.

Where several pumps serve a common rising main, the discharge and hence the friction loss will vary according to the number of pumps operating. The discharge of a given pump will be higher when it is working alone than when additional pumps are running, as the friction loss will be less. The difference may be quite large if the rising main is long and the static lift is relatively low.

Decisions on numbers and duties of pumps cannot be made in isolation from decisions on the size of rising main and since, as will be seen later, sump capacity depends on the discharge of individual pumps, this factor has also to be considered. Overall design therefore tends to be a process of compromise between the three elements: sump, pumping machinery and rising main.

5.2 Rising main

The rising main must be of such a diameter that solids deposited while the pumps are stopped will be scoured out when the pumps are working. A velocity of at least 1.2 m/s is commonly considered to be desirable.

A main of small diameter will be low in capital cost but it may be cheaper in the long run to use a larger diameter to reduce that part of the energy cost which is attributable to friction losses. Depending on the length, on the interest rates at which capital is borrowed and on electricity charges there will be some of diameter which minimises total annual cost. While it is worthwhile and customary to analyse costs of alternative sizes for aqueducts and hydro-power penstocks, where there are analogous problems, it is rare for this to be done in sewage pumping. In some cases the economic size would be so great that velocities would be too low. In others the potential savings are only of the same order as the range of error in estimating prices. Nevertheless, for long mains carrying large flows investigation will be worthwhile.

Where several pumps operate incrementally, the velocity will depend on the number of pumps working. If the main is small enough to have a scouring velocity with only one pump working, the friction losses might be unacceptably large when all pumps are working, and a larger main may have to be used. Provided that occasions when more than one pump is discharging and velocities are high are fairly frequent little trouble should be experienced. Thus with pumps of nominal capacity $1\frac{1}{2} \times$ DWF, the velocity might be about 0.6 m/s with one pump alone. Even in dry weather two pumps would have to operate for some part of each day giving a velocity of 1.2 m/s. One or more additional pumps would be needed for wet weather giving velocities up to 2.4 m/s at 6 × DWF.

If the pumping distance is not too great, a pair of rising mains might be considered with provision for changing the duties of the mains, or of the pumps serving them, every few days.

Rising mains are most commonly of cast-iron with spigot-and-socket joints. Asbestos cement pressure pipes and steel pipes are alternatives. Wherever there is a summit in longitudinal profile, air valves are needed. Provision is needed for emptying occasionally and pipework in the station is sometimes arranged in such a way that as much as possible of the contents of the main can be allowed to gravitate to the suction well. Hatch boxes are provided at intervals for rodding to remove blockages.

On large schemes there is a risk of water-hammer bursting the main when the pump stops and the reflux valve at the pump delivery closes. This risk can be avoided by installing a pressure-relief valve or an air vessel.

5.3 Number and duty of pumps

No general rules can be offered as to the choice of number of pumps and the division of the total flow among them. Standby units are always provided and there are obvious advantages in having several identical sets.

For a foul sewage pumping station on a separate sewerage system the maximum flow will be about 4 × DWF, which can be divided between two pumps working in parallel, a third set being provided as standby. On a large system the maximum flow might be divided among three or four pumps with two standby sets.

On combined systems, unless discharge is to coastal water without treatment other than screening, every effort will be made to site the pumping station so that a storm overflow can be installed upstream. The station will then be designed to take the maximum through-flow from the overflow chamber. This may be higher than the 'setting' or first spill through-flow and needs to be estimated by considering the hydraulics of the overflow chamber. To avoid long intervals between pumping in dry weather, units of somewhat less than 2 × DWF capacity will be needed, four or five of them working in parallel at maximum flow with two standby units.

Where a very wide range of flow has to be covered, as in pumping combined sewage where there is no overflow, or in pumping the spill from an overflow, consideration may have to be given to using pumps of two sizes. The smaller size pumps cater for the low flows and the larger size pumps cater for flow too great for the small pumps. In some cases the large pumps deal with the whole of the maximum flow, the small ones working only at low flows. When two sizes of pumps are installed standby units for the smaller pumps may not be considered necessary.

Since several different makers produce models covering the most frequently needed ranges of duties, a preliminary design is produced first without reference to any particular models, though quotations may be obtained from a few makers for inclusion in the estimate. After approval of the scheme has been obtained, tenders will be invited for the machinery. The successful tenderer becomes a nominated subcontractor and supplies firm details on which the final design is based. To avoid delay, the machinery and pipework may be ordered before the main contract is let.

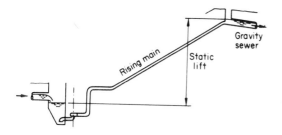

Fig. 5.2   Static lift for pumping schemes

The objects of the preliminary design calculations are to assess approximately the head, discharge and power requirement of each pump in relation to the diameter of rising main chosen. The static lift, see Fig. 5.2, will depend on the levels between which sewage is to be raised. The static lift will vary according to sump-water level, but the pumps will be adequate if they can maintain the design discharge when the sump is full to starting level. At stop level, the discharge will be smaller and some designers prefer to use the maximum rather than the minimum static lift to provide a small safety margin.

To the static lift must be added friction losses. These are assessed most readily by converting all bends, valves, etc., into equivalent lengths of straight pipe, see Table 6.3, before applying a pipe-friction formula. Where several pumps work in parallel, the discharges in the suction and delivery lines of the pumps will differ from those in the manifold system connecting the discharge lines to the rising main. The discharges in the suctions, deliveries and manifold cannot be determined precisely without pump characteristic (head: discharge) curves, so in the preliminary calculations the nominal intended discharge is used to assess the losses. The discharge for the rising main is of course the total station output.

Strictly, the kinetic energy head changes between the sump, where kinetic energy is zero, and the discharge should be taken into account but these have little effect on the total head as the velocities are quite low. If the suction pipes are long the kinetic energy change at the inlet might need to be considered to ensure that pressure is not too low at the pump. This is especially so where the pump is above water level.

Having calculated the total head for each pump for the case when all working pumps are delivering, reference to makers' catalogues will enable a provisional choice of pump size to be made and the dimensions given will be used to prepare a preliminary layout.

At the stage where a final choice of pumps is made, characteristic curves obtained from the maker enable a more exact analysis to be made of station performance according to the number of pumps running. This is illustrated in Fig. 5.3 for a case where two identical pumps in parallel serve the same main. Curve A is the characteristic for one pump running alone. Curve B is

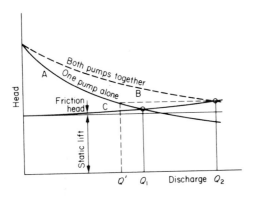

Q_1 output with one pump running alone

Q_2 output with both pumps running

Q' output of each pump when both running $(\frac{1}{2}Q_2)$

Fig. 5.3   Head and discharge for two pumps delivering through a common rising main

the head-versus-discharge relationship for both pumps running and is obtained by doubling the discharges of curve A. Curve C is static lift plus friction losses in the system, versus discharge. The intersection of curves B and C gives the discharge when both pumps are running. The intersection of curve C with curve A gives the discharge of one pump working alone. It will be observed that this is more than half the discharge for both pumps working together. Strictly, curve C will be higher for the single pump as all the flow will be passing through the suction and discharge lines of one pump.

The same principle can be used to find discharges for stations with several pumps of different sizes when any combination of pumps is working. It is important to check that the total discharge of the station is adequate, but it is rarely necessary to work out discharges precisely for times when fewer pumps are working.

Sewage pumps are nearly always coupled directly to a.c. motors and therefore run at speeds of somewhat less than 750, 1000, 1500 or 3000 rev/min (for 50 cycles/sec supply) according to the design and number of poles of the motor. It is rare to use V-belt drives or gears to run pumps at intermediate speeds, so the question of finding a speed to suit a given model to given head and discharge seldom arises.

5.4 Motors and starting gear

As mentioned above, a.c. motors fed from the public supply are now by far the commonest form of motive power. The electricity supply authority needs to be consulted at an early stage to see what sizes of motor and types of motor and starting gear are feasible at the site. If motors larger than about 8 kW are required, the starting current will probably have to be limited by suitable choice of motors and starters. The authority's kVA charge may also influence the choice of equipment. These factors may have to be taken into account when deciding on the number and size of pumps. In the larger pumping stations the supply authority may require a separate compartment to be provided for their own equipment.

The squirrel-cage induction motor is the simplest and cheapest machine but takes a high starting current and runs up to speed very quickly. Direct on-line starting is feasible only for the smaller sizes. Above about 8 kW an automatic star-delta starter would usually be required unless the supply were from the high-voltage network through a transformer. The starting current can be further limited by using an auto-transformer starter and this has the additional advantage of causing the motor to run up to speed and to stop more gradually, which may be desirable when pumping against high heads. Slip-ring or wound rotor induction motors would be an alternative choice to the squirrel-cage type in these cases. Synchronous motors are expensive in first cost but may be cheaper to run if the kVA charge is high.

Starters are designed to perform a chosen number of starts per hour without overheating. If the frequency of starting is excessive, the starter will not operate until the temperature has fallen. This affects the capacity of suction well needed. It is usual to provide sufficient capacity to limit the starting rate to $7\frac{1}{2}$ or to 15 starts per hour but starters are commonly used which will operate 30 or 40 times per hour. Starters capable of more frequent operation would be unduly expensive.

So that pumping stations can work with a minimum of attendance, starters are operated by water-level registering equipment in the suction well. This may consist of a float attached to a counter-weighted wire or to a rod. Trips on the wire or rod operate the starters at prearranged levels. An alternative is to use electrodes which cause relays to operate the starters when the water level rises above them or falls below them.

Switchgear is arranged so that any pump can be reserved for standby duty and so that the standby pump is automatically brought into operation if another pump fails to start.

5.5 Suction-well capacity

The flow of sewage varies continuously but constant speed pumps have fixed output and therefore run intermittently. Balancing capacity is provided by the suction well. As indicated above, sufficient capacity must be provided to ensure that the pumps are not required to start too frequently.

With a single pump, the cycle time T is given by:

$$T = \frac{c}{q} + \frac{c}{Q - q} \text{ minutes}$$

in which q = rate of flow of sewage into well, volume per minute,
$\qquad$ Q = pumping rate, volume per minute,
and $\qquad$ c = capacity of well between start and stop levels.
The start level will be such that the pump starts before the flow in the incoming sewer is backed up and is usually a few inches below the invert of the sewer. The stop level is near the bottom of the well but sufficiently above the suction pipe to avoid air entrainment.

For a given capacity c there will be a critical value of q which causes T to be a minimum, and the starting rate to be most frequent. This value of q is found by putting dq/dT equal to zero which gives

$$q = \tfrac{1}{2}Q \text{ for minimum } T$$

Consequently, the minimum of T is $4c/Q$.

If T is to be not less than $60/N$, in which N is the maximum number of starts per hour,

$$\frac{4c}{Q} \nless \frac{60}{N}$$

and $\qquad$ $c \nless \dfrac{15Q}{N}$

For $N = 7\tfrac{1}{2}$ per hour this gives the simple rule that c should be at least equal to the volume pumped in two minutes. No account has been taken of the time occupied in starting and stopping which will lengthen the cycle time and reduce the frequency slightly.

Where more than one working pump is installed, starting levels are arranged so that pumps are started successively as the flow increases. Thus if the flow is greater than the output of the first pump, water level will continue to rise after the pump has started and a second pump will be brought into operation when some chosen higher level is reached. Commonly both pumps continue to run until the well is emptied to the stop level which is common to both pumps, though if the stopping of both pumps simultaneously would cause undesirable water-hammer effects in the rising main it may be necessary to stop the second pump at a higher level than the first.

The suction-well capacity between start and stop levels of the first pump will be as given above. The determination of the capacity between the successive starting levels of additional pumps presents a far more difficult problem. This is further complicated because even if the pumps are identical they will not provide equal increments of output when they discharge through the same rising main as friction losses and consequently total head will increase with the pumping rate.

It is probably easier to find critical input rates by trial and then to use them to determine capacities rather than to attempt mathematical analysis. For two identical pumps with only a slight decrease in individual output when both are running, and a common stop level, the critical input rate is about $1\tfrac{1}{3}$ times the output of a single pump and the capacity needed between the two start levels is about 0.36 of 15 Q_2/N where Q_2 is the incremental output when the second pump is added. The total capacity needed is therefore appreciably less than one minute's discharge of both pumps for a starting rate of 15 per hour.

Starting levels have in any case to be separated by about 150 mm depth to avoid hunting effects from surface waves.

It is sometimes desirable to provide sufficient suction-well capacity to accommodate the contents of the rising main, so that repairs can be carried out, or to provide storage capacity for emergencies. These factors may override considerations based on pump starting frequency. In small stations, on the other hand, it may be necessary to limit pump and suction-well capacity so that the occasions of pumping in dry weather are not too infrequent.

5.6 Station layout

A typical design is shown in Fig. 5.4. The following features should be noted.

1 Flanged pipework is used inside the station. An attempt is made to use standard fittings and lengths as far as possible. One or two special make-up pieces are usually needed. For a few chosen joints, Johnson couplings or other adjustable joints are used instead of flanges so that the pipework can be easily dismantled and re-erected when repairs are needed. All pipes should be supported from the floor or walls so that joints are relieved of stress. This is especially important when flexible joints are used. Not all the supports are shown on the figure.

After a final choice of pumps has been made and dimensions obtained from the manufacturer, a separate pipework drawing is prepared with a schedule of all the components. In working out the dimensions, the thickness of joint rings should be taken into account.

The lengths of suction piping passing through the wall have 'puddle' flanges and are normally cast-in when the walls are concreted. The rising main may also be cast-in, the provision of adjustable joints being relied upon to make erection of the pipework possible. Alternatively, a fairly large hole may be left to receive the rising main, concreting being completed after the pipework is fixed. Whichever method is used, the designer must assure himself that the pipes can be put together without difficulty.

2 Each pump has a sluice (or gate) valve on both suction and discharge branches for isolation. A reflux (non-return) valve is provided on each pump discharge line. A sluice valve is provided on the rising main at the point where it leaves the station. In some cases additional pipework is provided to enable the rising main to be drained into the sump.

3 A small diameter air release pipe is needed from the top of each pump casing. These pipes are taken through to the wet well above water level. Plugged crosses are used instead of bends so that the pipes can be rodded.

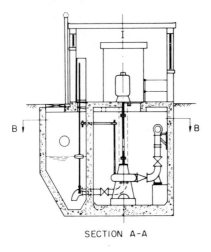

SECTION A-A

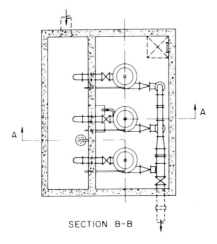

SECTION B-B

Fig. 5.4 Sewage pumping station

4 Water accumulating in the dry-well from pump glands and condensation is collected in a shallow sump. Pipes of small diameter, with valves, branching from one or more of the pump suctions enable this water to be removed.

5 The floor of the superstructure should have holes covered by gratings large enough to enable pumps to be removed for repair or replacement. A beam at ceiling level carries chain blocks for lifting.

5.7 Screw-lifts

Where liquid has to be raised between low-level and high-level channels which are close together on plan, the Archimedean screw-lift has a number of advantages. This device is in well-established use for land drainage in the Netherlands and is becoming the usual way of lifting sewage and activated sludge at treatment works.

The form of installation is shown in Fig. 5.5. The screw is formed from steel plate welded to a steel tube. The lower end runs in a bearing with a cast-iron, bullet-shaped housing. A thrust bearing is provided at the upper end of the screw. Electrical drive is from a motor and gear box on the axis of the screw or from a motor with horizontal axis at right-angles to the screw via V-belt and gear box.

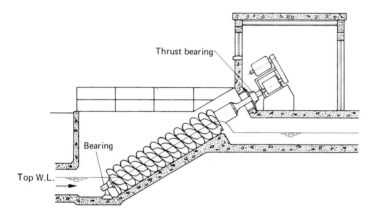

Fig. 5.5 Screw-lift installation (*Courtesy Simon-Hartley Ltd*)

The trough in which the screw rotates may be of steel plate or concrete. Concrete troughs can be screeded by the action of the screw.

To prevent the contents of the upper channel draining back when the pump is stopped, a non-return flap is provided. Alternatively, the screw may discharge into a channel below the lip at the upper end of the screw trough.

A screw of 500 mm diameter running at about 80 rev/min will deliver about 0.05 m³/s. Output increases roughly with diameter raised to a power somewhat less than 2.5, the larger screws operating at lower speeds. A 1.5 m screw would deliver about 0.7 m³/s at a speed of about 40 rev/min. Output depends also on the ratio of tube to screw diameter and on the inclination.

The maximum lift to which a screw can be applied depends on the need to avoid excessive bending of the tube and therefore on the diameter and ultimately on the desired maximum output. The maximum lift is therefore least for the lowest output.

The mechanical efficiency of a screw-lift when working at full design output is about 75%. When the rate of flow to the screw decreases, water level in the approach channel and the depth of submergence of the lower end of the screw are reduced but efficiency is still above 60% when the flow is only one quarter of the maximum design output. The efficiencies quoted are the proportion of screw-shaft power which is converted to water power. The efficiency of conversion of electrical power to screw-shaft power is about 75%.

As a screw-lift will deliver any rate of flow up to its maximum output a storage sump as needed for fixed-output centrifugal pumps is unnecessary. Where low flows are expected for long periods, however, or where the flow is intermittent, it may be desirable to provide float- or electrode-operated starting and stopping gear.

Where a wide range of flow has to be dealt with, two or more screw-lifts are installed in parallel.

Owing to its relatively simple construction and slow speed of rotation, a screw-lift can be used for unscreened sewage and is especially useful where the static lift would be too low for a centrifugal pump. In a flat district, it may be economical to use low-lift screws at several successive points in the system, rather than to run gravity sewers at increasing depths of cover to a single high-lift centrifugal pumping station.

5.8 Ejectors

When the flow is less than about 10 litre/s a pump would have to be so small that blockage would be frequent. A sewage ejector is then used, simple and reasonably trouble-free in operation but, mechanically, less efficient than a pump.

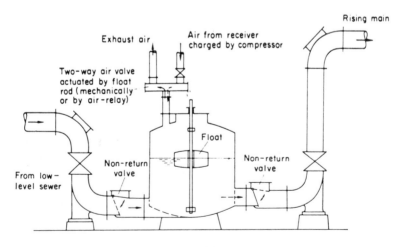

Fig. 5.6   Pneumatic ejector (diagrammatic)

The principle of operation of the ejector is shown in Fig. 5.6, and is similar to that of a reciprocating pump except that air pressure is used instead of a piston. The ejector body fills under gravity, the outlet valve remaining closed due to the pressure of water in the delivery pipe. Air is forced out of the body to the atmosphere. When the body is nearly full, the float trips a valve which admits compressed air to the body, closing the exhaust branch at the same time. The sewage inlet valve closes under internal pressure and the outlet valve opens, sewage being forced along the discharge pipe. When the body is empty the float reverses the air valves and the cycle is repeated. In some models the air-valves are operated with the aid of a balance weight while in others compressed air is used to activate them. The operating cycle usually takes about one minute so that the capacity of the body in litres gives the discharge in litres per minute. Capacities range from 45 to 680 litres. There is some advantage in having more than one unit and occasionally it is justifiable to provide standby capacity.

An air receiver between the compressor and the ejectors reduces the frequency of starting of the compressor motor and also the size of compressor needed. The compressor is then controlled automatically by pressure switches on the air receiver.

5.9 Inverted siphons

Where a sewer has to cross a cutting or river, headroom requirements sometimes preclude a bridge and the sewer must descend to a lower level to pass under the formation or bed. If head has to be conserved, the sewer must rise beyond the obstruction. The low-level pipes are then continuously full and under pressure. These pipes are not siphons in the true sense but the term 'inverted siphon' is commonly used to describe them.

As some silting is inevitable at low flows, the velocity at high flow must be sufficient to clear the accumulations. A mean velocity of about 1 m/s at maximum flow is usually considered to be the minimum for a nearly horizontal pipe but higher velocities are needed for pipes which rise in the direction of flow, increasing to 1.5 m/s in vertical pipes.

Wherever possible inverted siphons should be designed to run at velocities as high as the available head allows. The higher the design velocities, the smaller the chance of occasional blockage. Depending on the velocities which can be achieved, and hence upon the head available, the following expedients for minimising silting can be considered.

(1) Multiple pipes

A 'primary' pipe with a carrying capacity of $1\frac{1}{2}$ to $2 \times$ DWF runs continuously. Scouring velocities will be achieved for a few hours each day in the pipe. When the flow exceeds the capacity a weir in the forebay is overtopped and the surplus passes along a secondary pipe of about the same capacity as the primary pipe. When the first two pipes are carrying their maximum discharge, a weir on the opposite side of the forebay is overtopped and a tertiary pipe carries the excess flow. On a completely separate system, two pipes may be sufficient. Even if a single pipe can be considered reasonably satisfactory it is considered prudent to provide a standby pipe in case the working pipe becomes blocked. For a combined sewer, three pipes are needed, the tertiary being much larger than the other pipes. The tertiary pipe will have a very small chance of becoming blocked and will often be sufficiently large to carry the whole flow safely should blockages occur in the smaller pipes.

(2) Provisions for emptying and cleaning

A scour branch may be placed at the lowest point. The chamber in which the scour branches are accommodated needs to be designed so that the contents of the pipes can be pumped to the tailbay. Penstocks or stop-logs are needed at each end of each pipe. A hatchbox or removable section of pipe enables rodding to be carried out. The scour branches are sometimes arranged so that the contents of one pipe can be pumped along another. The forebay penstocks or stop-logs can be closed to back up the flow for a short time and then quickly opened to create a scouring flow.

(3) River scour

Occasionally, where the pipes pass beneath a river and the downstream sewer is below river level, it is possible to provide branches enabling river-water to be flushed along the pipes. If the upstream sewer is below river level, penstocks are needed in the forebay to prevent surcharge of the sewer by river-water.

(4) Storm overflow

The range of discharge to be accommodated can be reduced if a storm overflow can be installed upstream. In some circumstances the overflow can form part of the forebay, the siphon pipe or pipes acting as the module.

(5) Screens and detritus pits

Owing to the disposal problem, the installation of screening and grit removal plant (see Chapter 9) upstream of a siphon is rarely feasible except where such plant is virtually part of a treatment works the remainder of which is on the downstream side of the obstruction.

(6) Pumping

The installation of a pumping system is nearly always a possible alternative to an inverted siphon though it should be remembered that rising mains have

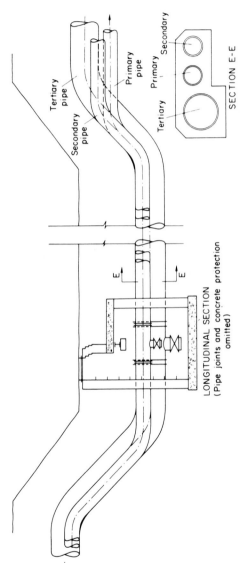

Fig. 5.7 Inverted siphon. (See Fig. 5.8 for details of chambers at head and tail)

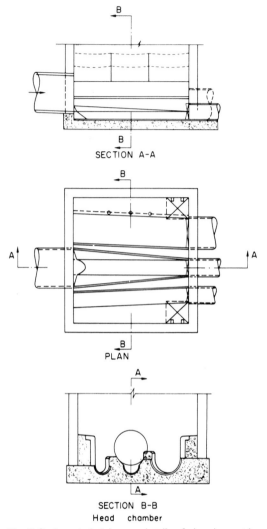

Fig. 5.8   Inverted siphon — details of chambers at head and tail

some problems in common with siphon pipes. Screw-lifts (Section 5.7) could perhaps be used in place of the rising sections of inverted siphon pipes. The downstream sewer could then be higher than the upstream sewer if necessary.

5.10 Design of inverted siphons

This section should be read in conjunction with Chapters 6 and 7.

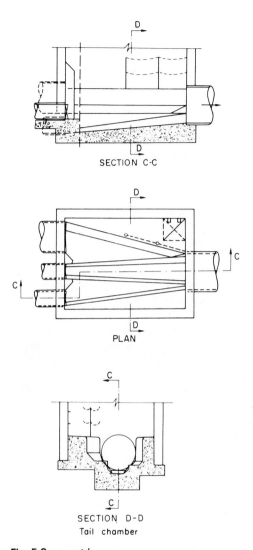

SECTION C-C

PLAN

SECTION D-D
Tail chamber

Fig. 5.8 — *contd*

The mode of operation and hydraulic design of a three-pipe inverted siphon will be described in relation to Figs 5.7 and 5.8. Ease of construction and ease of exact analysis tend to be incompatible. The details of the design shown in the figures are strongly influenced by considerations of simplicity of construction. Some possible alternatives will be mentioned.

The gravity sewer downstream of an inverted siphon will almost always be of mild slope for the discharges to be considered and will therefore act as a control for the whole siphon system. In most cases the sewer will be sufficiently long for the depth of flow at its upper end to be taken as normal depth. The depth–discharge relationship for the head of the sewer can then be obtained from Fig. 6.4.

The nominal capacity for the primary pipe will usually be between 1 and 2 × DWF. Several trials of the whole design will need to be made to find the best way to proportion the flow between the pipes. In a particular trial, the chosen nominal capacity together with the desired maximum velocity will determine the diameter of the pipe. The velocity will need to be between 0.9 and $1\frac{1}{2}$ m/s depending on the slope of the rising pipe. If head is to be conserved it may be desirable to use a diameter giving about 1 m/s for the falling and flat portions and to taper down to a smaller diameter giving a higher velocity in the rising pipe.

The head loss through the primary pipe when carrying its nominal capacity can be determined from a pipe-friction formula. It is most convenient to allow for bend losses by adding equivalent lengths of pipe (Table 6.3). Fairly high roughness should be taken to ensure that there is a margin for silting.

Flow is intended to spill over the left-hand weir on section BB when the primary pipe has reached its nominal carrying capacity. A first approximation to the required level of this weir will be the level of flow leaving the tail-chamber plus the head loss in the primary pipe. If it were necessary to ensure spill at a precisely specified flow, surface profiles would have to be calculated for the central channels of the head- and tail-chambers, estimates having been made for the head losses and transformations at their exits. Calculation of surface profiles will often be considered to be an unnecessary refinement.

In the design shown the sill on the secondary-pipe side of the tail-chamber channel is at the level of flow in the downstream sewer corresponding to primary pipe capacity. The sill on the opposite side is at a level corresponding to flow at the combined nominal capacities of the primary and secondary pipes. The central channel confines low flow. The outlet of the primary pipe is entirely below the lower sill and the channel bed has an adverse slope. It would be possible to raise the outlet of the primary pipe to flatten this slope. The backwater curve would then extend into the pipe. Depth of flow in the channel would be reduced and head loss would be slightly increased. Precast channel of constant diameter could be used in the tail-chamber but might increase head loss.

The centre channel of the head-chamber is almost level and its invert is below that of the incoming sewer. The difference in invert level could be reduced at the expense of either backing up the flow in the sewer or having shallow flow in the channel and in the entrance of the primary pipe at first spill. Neither disadvantage would be serious, though the former might lead to a need for longer weirs.

The left-hand weir of section BB needs to be long enough to spill the nominal capacity of the secondary pipe without creating an undesirable degree of backing-up in the sewer. In most cases the secondary pipe will have the same order of capacity as the primary pipe.

When the flow is approaching the combined nominal capacities of the two pipes, the flows in the separate pipes and over the weir are interrelated and it is not possible to make exact calculations. The water level in the downstream sewer will of course depend on the total discharge, and the tail-water levels for the primary and secondary pipes will be approximately the same. There will be a small difference as the primary flow passes along the tail-chamber channel while the secondary flow passes over the benching.

The water level in the head-chamber controls both the flow in the primary pipe and the flow over the weir. Lengthening the weir without changing any other dimensions would tend to increase the proportion of flow taken by the secondary pipe. The discharge of the primary pipe could conceivably decrease as total flow increased with a long weir, the tail-water rising faster than the head-water. A stage can be reached where there is sufficient flow through the secondary pipe for part of the weir to be drowned. The surface profiles in both primary and secondary head-chamber channels will be curved. Several types of profile are possible alongside side-weirs and drowning adds further possibilities to those which have been investigated.

Fortunately great precision of analysis is seldom warranted. If the primary channel is fairly large and the entrance to the primary pipe is submerged, the broad-crested weir formula (Section 7.6) will enable a rough approximation to be obtained for the water level in the head-chamber. Using this formula, a length can be found for the weir such that the sewer is not unduly backed up when the flow is equal to the nominal combined capacity of the two pipes. It may be worthwhile to estimate the flow in each pipe at this stage, neglecting surface curvature, and to check that the weir is unlikely to be drowned.

The sill of the right-hand weir is at estimated water level for the flow when the first two pipes are carrying their nominal design discharges.

The diameter of the third pipe is chosen to give adequate velocities when carrying the excess flow. Water level in the tertiary channel at maximum flow will be roughly equal to water level in the tail-chamber plus the head loss through the pipe. Energy transformations where there is free surface flow will make the true level somewhat higher. The water level should be below the weir sill. If there appears to be a risk of more than trivial drowning of the weir by the flow in the tertiary pipe, it is advisable to try a new design with a different allocation of flow between the pipes and different diameters and levels.

The weir needs to be long enough to discharge the excess flow without surcharging the sewer. The broad-crested weir formula may again be used for a rough estimate but if surcharging would have serious consequences a more accurate analysis might have been attempted. Complexities similar to those of the first weir arise.

Making both weirs the same length as shown has some attractions and can be done by making a compromise which may involve altering the sill levels. The pipe diameters will not need to be altered provided some alterations to the flow allocation can be tolerated.

The tail-chamber benchings over which flow passes through the secondary and tertiary pipes are flat in the illustration. They could be sloped downwards to the centre channel if the pipes were raised. The overall head loss would be slightly increased. On the other hand it might be found necessary to lower the outer pipes, particularly the tertiary pipe, and to slope the benching *up* to the sill. Water would remain behind the sill after discharge ceased but it would be possible to redesign the pipes and the head-chamber for a smaller overall head loss.

A possible improvement to the design shown would be to move the outer pipes nearer to the centre. The primary pipe is sufficiently low for the tertiary pipe to be moved about half a diameter and the secondary pipe could be moved to nest between the primary and tertiary pipes. The area of benching would be considerably reduced. Some difficult but not insuperable problems of draughtsmanship would arise in detailing the pipework.

In deciding on the nominal capacity of the primary pipe, the minimum flow should be considered. The minimum velocity in the primary pipe should not fall much below 0.3 m/s so the tendency is to make the nominal capacity of the primary pipe a smaller multiple of DWF where the minimum of flow is relatively low. The daily dry-weather peak flow is of interest in that the notional capacity of the first two pipes should not exceed it by very much. If the minimum flow is not too low, it may be possible to evolve a two-pipe scheme, the first pipe carrying flows up to about the daily peak.

It will be realised that the design of inverted siphons is something of an art. Many of the hydraulic elements would be difficult to analyse even in isolation, but direct calculation is made even more difficult by their interrelationships. Furthermore, the design requirements cannot be precisely defined since they represent a compromise between the conflicting needs to minimise head loss, to avoid upstream surcharge at maximum flow and to achieve self-cleansing velocities for as much of the time as possible. Several alternatives should always be explored. Rough calculations are sufficient to distinguish the promising versions. More refined calculations may be used to make small adjustments to the final design.

Model testing may be worthwhile but an attempt will have to be made to simulate the correct resistance in the siphon pipes if the results are to be reliable. For example, the length of the flat portions of the pipes could be increased to give a head loss corresponding to that calculated for the prototype. An adequate length of the downstream sewer will be needed with adjustable slope so that the calculated tail-water levels can be obtained. A short length of pipe discharging into a channel with an adjustable tail gate is an alternative.

References

British Pump Manufacturers Assocn, *Pump Users' Handbook*, F. Pollak (Ed.), Trade and Technical Press, 1970.
Anderson, H.H., *Centrifugal Pumps*, Trade and Technical Press, 2nd. Ed., 1972.
Addison, H., *Centrifugal and other rotodynamic Pumps*, Chapman and Hall, 1966.

6

Flow in pipes and channels

Notation

b	bottom width of rectangular channel
d	diameter of pipe
f	Darcy–Weisbach friction factor, $2gdi/4V^2$ or $2gmi/V^2$
g	acceleration due to gravity
h_f	energy-head loss
i	energy-head loss per unit length of conduit
k_s	linear measure of equivalent sand roughness
m	hydraulic mean depth, A/P
n	measure of roughness in Manning formula
y	depth of flow
A	cross-sectional area of flow
C	coefficients in Chézy formula, $V = C\sqrt{mi}$
C', C''	coefficients in formulae other than Chézy
P	wetted perimeter
Q	rate of flow
V	mean velocity, Q/A
$\mathbf{D}$	diameter parameter, d/k_s
$\mathbf{S}$	gradient parameter, $2gik_s^3/\nu^2$
$\mathbf{R}$	hydraulic mean depth parameter, m/k_s
$\mathbf{Q}$	discharge parameter $(4\pi m/P)\,(Q/k_s\nu)$
$\mathbf{V}$	velocity parameter Vk_s/ν
$\mathcal{R}$	pipe-flow Reynolds number
$_0$ (suffix)	applied to A, m, Q or V to signify values for full-bore flow exponent of d in an exponential pipe-friction formula $(\alpha = 3\beta - 1)$
α	in Section 6.4 only, Coriolis velocity distribution coefficient
γ	exponent of d/k_s in an exponential pipe-friction formula
δ	exponent of ν in an exponential pipe-friction formula $(\delta = 2\beta - 1)$
ν	kinematic viscosity

6.1 Frictional resistance to flow in pipes

When a fluid is moving along a pipe, work is being done against the frictional resistance of the wall. Consequently the energy of the fluid decreases as it moves along. The energy loss per unit length can be equated to the work done against friction per unit length as in the Darcy–Weisbach formula:

$$i = \frac{fV^2}{2gm} \tag{6.1}$$

For a circular pipe with full-bore flow the hydraulic mean depth, m, is $d/4$ and the formula becomes

$$i = \frac{4fV^2}{2gd} \tag{6.2}$$

Since the work to be described was done on circular pipes, this version will be used for the present. Non-circular sections are discussed later.

The friction factor f is dimensionless and is a function of the pipe-flow Reynolds number Vd/v and of a dimensionless measure of the roughness of the wall. The relationships between f and these parameters have been fully explored and are shown in Fig. 6.1. This is a graph of f against $\mathscr{R}$, logarithmic scales being used for both axes. The formulae on which the graph is based are listed in Table 6.1.

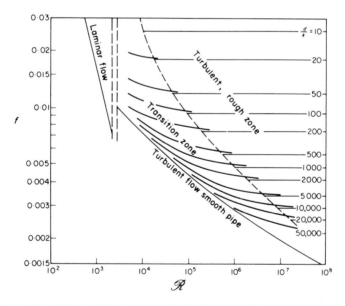

Fig. 6.1   Relationships between friction factor and Reynolds number in pipe flow

Table 6.1   Pipe Friction Formulae

Laminar flow	(1)	$f = \dfrac{16}{\mathscr{R}}$
Turbulent flow		
Smooth pipe	(2)	$\dfrac{1}{\sqrt{f}} = 4 \cdot \log \dfrac{\mathscr{R}\sqrt{f}}{1.255}$
Rough pipe	(3)	$\dfrac{1}{\sqrt{f}} = 4 \cdot \log \left(3.7 \dfrac{d}{k_s} \right)$
Transition	(4)	$\dfrac{1}{\sqrt{f}} = -4 \cdot \log \left\{ \dfrac{k_s}{3.7d} + \dfrac{1.255}{\mathscr{R}\sqrt{f}} \right\}$

At Reynolds numbers below 2300 the relationship is given by (1). The flow is laminar and the friction stress at the pipe wall is transferred to the mass of fluid by viscosity alone. Mean velocity can be connected with diameter and gradient by mathematical reasoning, starting with the definition of viscosity which enables the variation of velocity over the cross-section to be deduced. Flow in this range is of no practical interest here.

At Reynolds numbers above 2300, conditions are unstable. Laminar flow can exist but is easily upset. At values above 3000 the bulk of the flow is turbulent but in relatively smooth pipes the flow near the wall remains laminar. The thickness of this laminar sub-layer decreases as the Reynolds number increases and the stage at which fully turbulent flow is reached depends on the size of the roughness elements on the wall in relation to the thickness of the laminar sub-layer.

In turbulent flow, the filaments of flow cease to be laminar. Instead, they roll up to form an ever-changing mass of eddies. Wall-friction stress is then transferred to the flow by means of these eddies and the situation is too complex to be analysed solely by mathematics. A combination of theoretical and experimental work has produced the equations (2), (3) and (4), Table 6.1.

The first one of these to be produced was (2). Several independent workers obtained results for what are now called hydraulically-smooth pipes. Tests carried out on such pipes over wide ranges of velocity and diameter and for fluids of different viscosities lie on a single curve when f is plotted against $\mathscr{R}$. Later Prandtl and von Kármán produced (2) to fit such results.

Rough pipes were experimented on by Nikuradse who used pipes artificially roughened with uniform sand grains and found that the ratio of pipe diameter to grain size was an important factor. For each value of this ratio, results gave a constant value for f, independent of $\mathscr{R}$, once a certain value of $\mathscr{R}$ was exceeded. The actual value of f depended on d/k_s as also did the value of $\mathscr{R}$ above which f became constant. Formula (3) was fitted to these results by Prandtl and von Kármán using similar reasoning to that used for the smooth-pipe formula. A criterion was also established for the boundary of the rough-pipe turbulent flow zone.

In the region between rough-pipe turbulent flow and smooth-pipe turbulent flow, Nikuradse's results differ from those for commercial pipes. A formula which fitted the latter was evolved by C.F. Colebrook and C.M. White. This is given as (4) in Table 6.1. It becomes identical with (2) when k_s is zero and approaches (3) as $\mathscr{R}$ increases.

The set of formulae in Table 6.1 and the chart in Fig. 6.1 cover nearly all cases of pipe flow which can exist, ranging from small capillaries to large tunnels, from thin fluids to highly viscous ones and covering a wide range of velocities and roughness values. They do not cover roughness of a regular nature in the transition zone and the Colebrook–White formula differs from Nikuradse's results. Further investigations by Morris[1] revealed that three different types of flow could be distinguished near rough walls and that, for a regular disposition of roughness elements, the spacing could be more significant than the height. Experiments by Ackers[2] on sewer pipes have shown that the effects of joints might be capable of being interpreted in the light of Morris's work but that this would lead to complications avoided by the use of the Colebrook–White formula which was found to be reasonably adequate. The question is an open one and it may be that some formula based on Morris's work will come into use for sewers in the future.

Figure 6.1 was drawn to illustrate the formulae in Table 6.1. The relationships between f, $\mathscr{R}$ and d/k_s are more familiarly shown on the Moody chart in which there is no discontinuity where the d/k_s lines cross the rough-turbulent zone boundary. Moody produced his graph by plotting results from a wide survey of pipe tests.

6.2 Use of the friction formulae in design calculations

The Moody chart or Fig. 6.1 can be used in design by calculating $\mathscr{R}$ and d/k_s so that these values can be taken on to the chart to find f which can then be used in the Darcy–Weisbach formula (6.2) to obtain i or h_f. In this procedure, d is needed at the outset to calculate both $\mathscr{R}$ and d/k_s.

The starting point for many practical design calculations is a rough idea of the velocity which will lead to an economic design in the particular application. This will usually enable the diameter to be narrowed down to not more than two commercially available sizes. Head loss is then worked out for each and a final choice is made.

As an alternative to the approximate value of f read from the graph, the logarithmic formulae can be used. The Colebrook-White formula is of universal application for turbulent flow since it gives acceptable results both for smooth pipes and in the rough-turbulent zone as well as in its own transitional domain. However, not being explicit in f, it has to be used iteratively.

The rough-turbulent formula gives f directly and, since many cases in wastewater engineering lie in this zone, it is always worth checking first. Figure 6.1 can be used for this or alternatively the value of $\mathscr{R}/(d/k_s)$ can be found and compared with 807, which, to a very close approximation, is the lower boundary of the rough-turbulent zone.

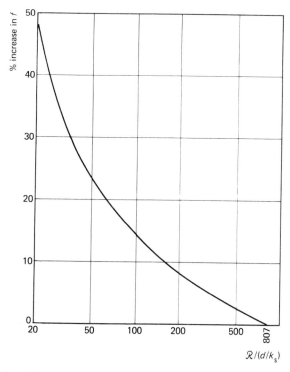

Fig. 6.2 Approximate percentage increase in friction factor *f* in Transition zone

If $\mathscr{R}/(d/k_s)$ is less than 807, it is still worth using the rough-turbulent zone formula to find *f* as this value can be used as a first trial in the transition formula. Alternatively the rough-turbulent value can be adjusted to obtain close approximations to transition values by the percentages shown in Fig. 6.2.

With ν as 1.15×10^{-6} m^{2}s^{-1} (water at 15°C) and k_s as 1.5 mm, cases are in the rough-turbulent zone at velocities above 0.6 ms^{-1}.

The Wallingford charts and tables[3,4], based exclusively on the transition formula, are frequently used in design offices.

6.3 Equivalent roughness values

The walls of pipes carrying sewage tend to be covered with a layer of slime of biological origin which grows quite readily when the pipe is new but then remains constant except for seasonal fluctuations. The thickness of the slime layer depends on the general velocity of flow, being greater for lower velocities, and depends also on the nature of the pipe wall. Material such as concrete acquires thicker layers than smoother material such as glazed clayware and uPVC.

It has become customary to quote values of k_s on a geometrically stepped scale: 0.15, 0.3, 0.6, 1.5, . . ., mm, the steps being in ratios of either 2 or $2\frac{1}{2}$. The comparison for a given material between a normal example and a poor or good example is considered to be one step up or down respectively, on the scale. For the range of parameters encountered in sewer or sewage works design a change of k_s by a multiple of 2 changes the calculated velocity by about 12%.

Recommended values relevant to sewers are shown in Table 6.2. Further data are given in the Hydraulics Research publications.[3,4]

Table 6.2   Equivalent sand-roughness values

Pipe material	Velocity (m/s)	k_s (mm) 0.15	0.3	0.6	1.5	3.0	6.0
Brickwork							
Well pointed					●	●	●
Needing pointing					up	to	30
Rising mains	1.1					●	
	1.3				●		
	1.5			●			
Slimed sewers	0.75						
Concrete, spun and							
vertically cast						●	●
Asbestos cement						●	●
Clay					●	●	
uPVC				●	●		
Slimed sewers	1.2						
Concrete					●	●	
Asbestos cement				●	●		
Clay			●	●			
uPVC		●	●				

In Table 6.2 the three values given for brickwork are for good, normal and poor examples. The values for rising mains are all quoted as 'poor'. For slimed sewers, two values are given. The upper one is suggested for design where a high value would be on the safe side. The lower value may be more realistic for the analysis of a flow in an existing system if no site value is available. Both high and low values may be regarded as 'normal' in the particular circumstances. The diameter of the pipe itself was used to deduce these values from observations[5] so no separate allowance for reduction in diameter need be made when using them for design.

Much less guidance is available for pipes and channels in sewage treat-ment works. Perhaps many of the pipes can be regarded as being similar to rising mains. Sewage works channels have fewer joints than pipes and if con-structed with smooth formwork may be no rougher than pipes. Also their surfaces may be treated to inhibit slime growth. On the other hand there is evidence[6] that k_s values are higher for partly full flow in pipes than for full-bore flow so, allowing one factor to balance another, values similar to those for pipes might be considered appropriate.

6.4 Non-circular sections and free-surface flow

For non-circular sections the Darcy–Weisbach formula is used in the form:

$$i = \frac{fV^2}{2gm} \tag{6.1}$$

Nearly all the fundamental work has been done on circular sections and the main evidence for the validity of this formula is that k_s values obtained with its use from tests of non-circular sections are not strikingly dissimilar from what might be expected for circular sections of the same material. However, slight discrepancies have been observed and it may be that hydraulic mean depth as a measure of shape and size would not be adequate for sections of peculiar shape. The discrepancies are not great for sections of practical shape and the formula is quite adequate for ordinary use.

Non-circular cross-sections of flow are most often encountered in open channels or pipes running partly full. These are compared with pipes running full under pressure in Fig. 6.3(a), (b). In both cases, i is the rate of loss of

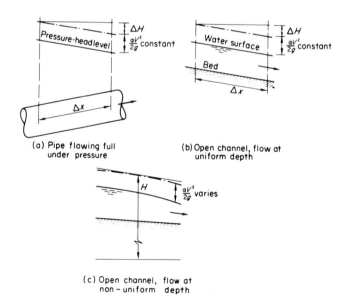

(a) Pipe flowing full under pressure

(b) Open channel, flow at uniform depth

(c) Open channel, flow at non-uniform depth

Fig. 6.3 Energy head in pipes and channels. In cases (a) and (b), $i = \Delta H/\Delta x$. In case (c), $i = dH/dx$, the rate of fall of the tangent to the total energy-head line, with respect to distance measured along the bed

total energy head per unit length of conduit. The difference between total energy head and potential energy head is $\alpha V^2/2g$. The Coriolis velocity distribution coefficient α has to be introduced because velocity varies across the section, while V is the mean velocity.

For a closed conduit, the potential energy-head line is above the top and may be observed at pressure tappings. For open-channel flow, the potential energy-head line is the water surface. If a pipe is running full-bore or channel flow is at uniform depth and the velocity distribution is the same at all sections, $\alpha V^2/2g$ is constant and the total and potential energy-head lines are parallel. They are straight since the rate of loss of energy is uniform.

In an open channel it is rare for the water surface to be parallel to the bed and the velocity energy head $\alpha V^2/2g$ varies along the length, so the surface (that is, the potential energy-head line) is not parallel to the total energy-head line. Because the depth is varying along the length, the rate of energy-head loss varies and the energy-head line is curved, Fig. 6.3(c). Depth varies in most channels because it is constrained at inlets, outlets and changes of section. Strictly, such constraints are absent only when the channel is infinitely long. In that case, the depth will be uniform along the length of the channel. This depth of flow, known as normal depth, will be such that the loss of total energy head per unit length through friction of the bed and walls will be equal to the loss of potential energy head per unit length. Total energy head, water surface and channel bed will all be straight and parallel to each other. The normal depth can be found from the Darcy–Weisbach formula. It is the depth at which

$$\frac{f(Q/A)^2}{2gm} \tag{6.3}$$

becomes equal to i, the hydraulic gradient, which in this case is the same as the bed gradient.

Where the flow is not at uniform depth, the Darcy–Weisbach formula relates to rate of fall of the total energy-head line in relation to channel distance only at a particular section. The analysis of non-uniform flow is dealt with in the next chapter.

6.5　Normal depth of flow

The calculation of normal depth is important because, even in a short channel, the water surface often approaches this depth asymptotically. Consequently it is needed in analysing non-uniform flow and, where analysis would not be warranted, it at least indicates the depth towards which the flow is tending.

In (6.3) above both A and m depend on y and for practical sections it is not possible to express the Darcy–Weisbach equation explicitly in terms of y. A further complication is that strictly f depends on y in an even more complex fashion.

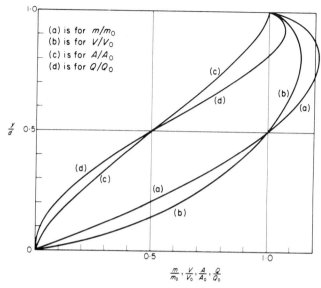

Fig. 6.4 Proportional depth relationships for partly full circular pipes based on $V \propto m^{2/3}$

For circular pipes running partly full it is relatively easy to find normal depth if an exponential formula is used. This is not strictly in accord with modern pipe-friction theory but gives results which are acceptable for most design purposes. Though the principles can be used with any exponential formula, it has usually been applied on the assumption that

$$V \propto m^{2/3}$$

as in the Manning formula. This assumption is reasonably correct for d/k_s-values greater than about 200 whether the flow is rough-turbulent, transitional or in a smooth pipe.

For pipes running full, variables will be given the subscript '$_0$'.

The ratio m/m_0 is a function of y/d and is plotted as curve (a) in Fig. 6.4. The ratio V/V_0 is the same as $(m/m_0)^{2/3}$ and its relationship with y/d is shown as curve (b). Curve (c) relates A/A_0 to y/d. Combining values from curves (b) and (c) the curve (d) for Q/Q_0 is obtained. This is the same as a curve of $Am^{2/3}/A_0 m_0^{2/3}$.

To find the normal depth of flow for a given discharge, diameter and gradient, the discharge Q_0 which the pipe would carry if it were running just full is found, using the diameter and gradient. The actual discharge Q is used to find Q/Q_0 and the graph gives the corresponding value of y/d.

From the curves it will be noticed that partly full circular sections have peculiar properties. At any given gradient, the pipe carries about 7% more flow at a depth of about 94% of diameter than it does when it is just full.

Strictly, the exponent of m is not constant but varies with d/k_s and hence with y. This can be seen by comparing the formula with the fundamental ones or more easily by comparing it with the Lamont formulae.

To see whether variation of the exponent has a significant effect from the practical point of view, curves could be drawn for different values of the exponent. This need not be done as Ackers[4] has gone to the root of the matter by producing a set of curves from the fundamental formulae. To do this it was found necessary to introduce a subsidiary variable θ. For rough-turbulent flow, θ is simply d/k_s. For smooth-pipe flow, θ depends on gradient, and for transitional flow both gradient and roughness are involved. Ackers' curves for proportional discharge show that within the practical range the effect of θ is slight. When a curve based on $Am^{2/3}/A_0 m_0^{2/3}$ is plotted on the same graph it is found to be a quite acceptable approximation to the true curve.

The method described for circular sections may be applied to rectangular and trapezoidal sections by using y/b in place of y/d in which b is the bed width, and by taking Q_0 as the discharge for flow at a depth equal to bed width. Figure 6.5 shows the proportional relationships for rectangular channels.

The procedure for finding the normal depth of flow in a rectangular channel for given discharge is: (1) use a friction formula or chart to find V_0 corresponding to $m = b/3$ (or $d = 4b/3$) and to the given gradient and rough-

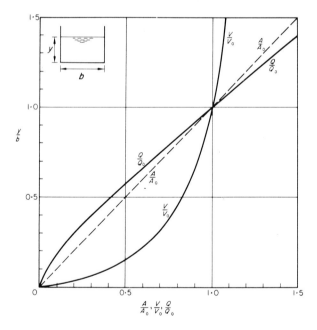

Fig. 6.5 Proportional depth relationships for rectangular open channels

ness; (2) calculate $Q_0 = b^2 V_0$; (3) calculate the ratio of the given discharge to Q_0; (4) use Q/Q_0 to find y/b from Fig. 6.5.

Since k-values show an apparent increase in free surface flow (see Section 6.3) it is wise to reduce Q_0 by a few per cent before calculating proportional discharge ratios for circular pipes. For rectangular sections this correction is necessary only when roughness values have been obtained from analogous pipe materials.

The curves in Figs 6.4 and 6.5 are based on roughness characteristics which remain the same irrespective of depth. In research on the sliming of sewers,[5] using free-surface flows varying on a daily cycle, slime grew non-uniformly round the periphery. The consequence was that the pipes carried their maximum discharge when running full, discharge having increased continuously with depth.

6.6 The Wallingford charts

In order to make modern pipe-friction formulae easier to use for design, Ackers devised the Wallingford chart[3,4] shown in Fig. 6.6.

Five variables occur in pipe-friction problems: V, d, i, k_s and ν. The

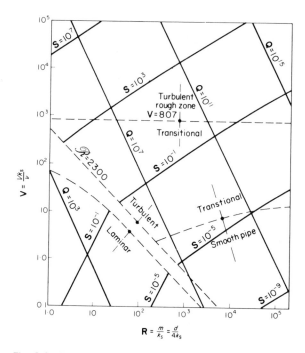

Fig. 6.6 The Wallingford chart (*Published by permission of the Controller of H.M. Stationery Office*)

designer is interested also in Q but this can be readily related to V and d. Strictly g is a variable but need not be considered as one here. In almost all design calculations k_s and ν are fixed at the outset. It is rare for a designer to need to calculate k_s from values of the other variables (though this has to be done when k_s-values are deduced from results of flow tests) and it is inconceivable that he would need to calculate ν. The problem is, therefore, to relate V, d and i for fixed values of k_s, ν and g.

In the Wallingford chart this is done by using three 'design parameters'. Each of these is related to one of the design variables and is rendered dimensionless by including k_s, ν or g.

The parameters are:

$$\mathbf{V} = \frac{Vk_s}{\nu}$$

$$\mathbf{D} = \frac{d}{k_s} \quad \text{or} \quad \mathbf{R} = \frac{m}{k_s} = \frac{\mathbf{D}}{4}$$

and $$\mathbf{S} = \frac{2gik_s^3}{\nu^2}$$

Once k_s, ν and g are fixed as at the start of a design calculation, the design parameters become directly proportional to the corresponding design variables.

The chart, see Fig. 6.6, has logarithmic scales of $\mathbf{V}$ and $\mathbf{R}$ for its axes. Lines of constant $\mathbf{S}$ are drawn upon it. The transition zone occupies the greater part of the chart and the Colebrook–White formula is used to plot the gradient parameter lines. In the laminar zone the lines are added for interest rather than for practical use.

The rough-zone boundary is approximated by $\mathbf{V} = 807$. Above this line the rough-turbulent formula could have been used, but the transition formula is close enough for the difference to be ignored and its use beyond the transition zone avoids discontinuities. As smooth-pipe flow is approached and $k_s = 0$, the scales in terms of design values become very widely spaced.

Most often the designer needs to find pipe diameter in relation to the discharge to be carried and the gradient available, so a design chart needs to have lines of equal discharge.

Lines of equal discharge can be easily added to a dimensional graph of V versus d for circular pipes running full, using

$$Q = \tfrac{1}{4}\pi d^2 V$$

The discharge lines are, of course, quite independent of the formula used for the energy gradient lines.

To complete the Wallingford chart, a dimensionless discharge parameter had to be evolved. The parameter is

$$\mathbf{Q} = 4\pi\frac{m}{P} \cdot \frac{Q}{\nu k_s}$$

which has the advantage of reducing to $\mathbf{Q} = Q/\nu k_s$ for circular pipes running full, the cases of most common occurrence. A graph of y/b or y/d versus $4\pi m/P$ for rectangular, trapezoidal and partly full circular sections enables cases of free-surface flow to be dealt with.

Expressing the discharge parameter in terms of the other design parameters:

$$\mathbf{Q} = 4\pi\mathbf{R}^2\mathbf{V} \quad \text{or} \quad \tfrac{1}{4}\pi\mathbf{D}^2\mathbf{V}$$

i.e. $\log \mathbf{V} = \log \mathbf{Q} - 2 \log \mathbf{R} - \log (4\pi)$

or $\log \mathbf{Q} - 2 \log \mathbf{D} - \log (\tfrac{1}{4}\pi)$

From these it can be seen that the $\mathbf{Q}$ lines on the chart have a downward slope of 2 to 1 and that they are at half the spacing of the $\mathbf{R}$ scale.

The Wallingford chart has been used to produce a set of dimensional charts[4] covering a wide selection of k-values. Tables based on the charts are available.[7]

The dimensional charts can be readily used for pipes running full but a trial-and-error method is needed for many problems involving free-surface flow. If the slight discrepancies between formulae in which $V \propto m^{2/3}$ and the logarithmic formulae are acceptable, these problems may be solved directly by using Fig. 6.4 or Fig. 6.5 in conjunction with values obtained from the Wallingford chart or table.

For circular sections, Q_0 is given directly by the dimensional chart.

For rectangular sections, Q_0 (for depth equal to width) is found for flow at depth b in which $m = b/3$ and $P/m = 9$. Alternatively, the mean velocity for $m = b/3$ can be found and multiplied by b^2 to obtain Q_0 as described previously.

The depth of flow is then found from Fig. 6.4 (circular pipes) or Fig. 6.5 (rectangular channels) using the ratio of the given discharge to Q_0, and remembering to reduce Q_0 slightly to allow for the apparent increase in roughness.

6.7 Energy losses at bends, etc.

These are usually regarded as being proportional to the kinetic energy head. The total energy loss through a pipe with bends, valves, etc., may be expressed as

$$H = \frac{4flV^2}{2gd} + k_1\frac{V^2}{2g} + k_2\frac{V^2}{2g} \tag{6.4}$$

where l = length of pipeline and k_1, k_2, k_3 are factors for the energy-head loss at the bends and valves. Since all terms involve $V^2/2g$:

$$H = \frac{V^2}{2g}\left(\frac{4fl}{d} + k_1 + k_2 + \ldots\right) \tag{6.5}$$

An alternative which is very useful in design calculations when the diameter is unknown, is to regard each fitting as being equivalent to the length of straight pipe which would have the same resistance.

Thus $\dfrac{kV^2}{2g} = \dfrac{4fl'\,V^2}{2gd}$

where l' = length of straight pipe with the same resistance as the fitting.

It is convenient to express l' in terms of numbers of diameters of straight pipe, N:

$$l' = Nd = \frac{kd}{4f}$$

$$N = k/4f$$

(6.5) now becomes:

$$H = \frac{4fl_c\,V^2}{2gd} \quad \text{or} \quad \frac{H}{l_c} = i = \frac{4fV^2}{2gd} \tag{6.6}$$

where $l_c = l + d(N_1 + N_2 + \ldots)$, the total length of equivalent straight pipe, and i is taken as the energy-head loss per unit length of equivalent straight pipe. The complications of fittings are thus removed at the beginning of the calculation.

Values of k and N obtained in different experiments vary a good deal. This is partly because valves, bends and other fittings differ in detailed design, partly because resistance varies with age and condition and partly because of interference between successive fittings in the same pipeline. Several

Table 6.3 Length of straight pipe with the same resistance expressed as number of diameters

Entrances of pipelines*	
Sharp-edged	20 to 25
Bell-mouthed	2
Foot valve and strainer	90 to 120
Sluice valve, fully open	5 to 7
Reflux valve of hinged flap type	40 to 80
90° bends	
Short	45 to 90
Medium	20 to 30
Long	10 to 20
Tees	
Flow straight through	10 to 20
Flow to branch	50 to 100

(45° bends have rather less than half the resistance of 90° bends of the same curvature. $22\frac{1}{2}$° bends have about half the resistance of 45° bends)

* There will be a conversion of potential energy to kinetic energy to be taken into account as well. See Fig. 4.16.

successive bends in the same pipeline may have less total resistance than the sum of the resistances of the same bends tested in isolation, particularly if the curvature is reversed.

Table 6.3 gives values of N from various sources. A more detailed table is given in the *Manual of British Water Engineering Practice*.[8] Since it is not possible to be precise, ranges are given so that a choice may be made according to the need to be conservative or otherwise.

In pumping systems, unless the rising main is very long, fittings cause the greater part of the total friction loss. Fittings losses are much less significant in inverted siphons.

6.8 Low velocities

If the velocity of flow in a sewer is too low, there will be a risk of solids being deposited. Sewers are therefore designed to flow at 'self-cleansing' velocities. The minimum self-cleansing velocity is commonly taken as 0.9 m/s, though larger sewers (diameters exceeding 0.75 m) are sometimes designed to flow at velocities down to 0.75 m/s. The values quoted are mean velocities for the pipe running 'just-full' and may be presumed to take account of the variations in discharge which commonly occur in sewers and of the variation in mean velocity with discharge. It may be noted (Fig. 6.4) that velocities exceeding the just-full value V_0 occur with discharges exceeding $\frac{1}{2}Q_0$ and that velocities do not fall much below V_0 except for quite small values of Q.

Scouring velocities can be estimated using a formula evolved by Camp[9] based on some earlier work of Shields:

$$V = \sqrt{\frac{2k}{f} g(s - 1)d} \qquad (6.7)$$

in which V = mean velocity at which particles on the bed begin to move
$\qquad\quad$ d = diameter of particle moved
$\qquad\quad$ s = density of particle/density of fluid
$\qquad\quad$ k = a factor varying from 0.04 to 0.8 according to the extent to which particles cohere.

It is interesting to compare this formula with the independently derived formula[10]:

$$V = 17v_s \qquad (6.8)$$

in which v_s = terminal-settling velocity of particles moved by mean-forward velocity V.

It is shown in Section 9.7 that

$$v_s = \sqrt{\frac{2g}{C} \cdot \frac{2d}{3} \cdot (s - 1)} \qquad (9.2)$$

Equations (9.2) and (6.8) imply values for k in (6.7) which lie between the limits quoted but which are somewhat higher than those which would normally be chosen when using (6.7).

According to (6.8), a mean velocity of 0.9 m/s will move grit particles (specific gravity 2.65) up to about $2\frac{1}{4}$ mm in diameter.

Scouring velocities are of interest in the design of grit channels (Chapter 9). The mean velocity commonly adopted in these is 0.3 m/s. According to (6.8), grit particles up to 0.25 mm in diameter would be retained in the flow.

Since mean design velocities are not allowed to fall below the minimum self-cleansing value, there is a minimum gradient for each diameter and class of pipe. Thus, 150 mm clayware and concrete pipes are not laid flatter than about 1 in 100. In theory smoother pipes can be laid at slightly flatter gradients. However, small pipes are so easily blocked that many engineers will not risk laying them at such flat gradients.

With sewers of large diameter, gradients for minimum self-cleansing velocities are so flat that there is little chance of maintaining the gradient accurately. Local variation in gradient and even backfalls are likely to develop due to uneven settlement. Whether the flatness of gradients should be limited for this reason is a matter for individual consideration. If the district is flat, extremely slack gradients may have to be accepted. Local variations from the design gradient are unlikely to affect the carrying capacity materially, but a need for regular de-silting may arise at some points in the system.

Where it is impossible to achieve a self-cleansing gradient in a small sewer, a diameter of pipe larger than that needed for the design flow is sometimes used. Thus a 225 mm pipe might be used in place of a 150 mm and laid as flat as perhaps 1 in 160. It will run about half-full with the design discharge, at a mean velocity about equal to its running-full velocity. The mean velocities for discharges less than the design value will not be so high as for the 150 mm pipe (see Fig. 6.4) and a justification for this practice is that there is space for deposits to accumulate at low flows without causing trouble. There will be a good chance of the deposits being scoured when the design discharge is passing.

Where larger sewers have to be laid at flat gradients, a non-circular cross-

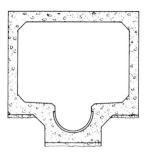

Fig. 6.7 Culverts of compound section

section has sometimes been resorted to. Concrete tubes or brick sewers of egg-shaped section were once fairly common. With sections of this shape mean velocities at low Q/Q_0 ratios are better than for circular sections. Their use is less common nowadays as designers have considered it unnecessary to concern themselves unduly with satisfying what is no more than a rough rule-of-thumb for minimum velocities.

Sections of U-shape (see Section 3.12) are triflingly better than circular ones from the point of view of velocities at small depths but sometimes have an advantage in cost.

For large surface-water culverts, compound sections (Fig. 6.7) have sometimes been used. The section is designed so that low flow is confined to the centre channel. Access for inspection and maintenance is improved.

References

1 Morris, H.M., A new concept of flow in rough conduits, *Proc. Am. Soc. Civ. Engrs*, **80**, No. 390, 1954.
2 Ackers, P., Hydraulic resistance of drainage conduits, *Proc. Inst. Civ. Engrs*, **19**, 307–336, July 1961.
3 Ackers, P., Resistance of fluids flowing in pipes and channels, *Hydraulics Research Paper* No. 1, H.M.S.O., 1958.
4 Hydraulics Research Wallingford, *Charts for the hydraulic design of channels and pipes*, H.M.S.O., 3rd Edn, 1983.
5 Perkins, J.A. and Gardiner, I.M., The hydraulic roughness of slimed sewers, *Proc. Inst. Civ. Engrs*, Pt. 2, 87–104, March 1985.
6 Ackers, P., Crickmore, M.J. and Holmes, D.W., Effects of use on the hydraulic resistance of drainage conduits, *Proc. Inst. Civ. Engrs*, **28**, 339–359, July 1964.
7 Hydraulics Research Wallingford, *Tables for the Hydraulic design of pipes and sewers*, H.M.S.O., 4th Edn, 1983.
8 Skeat, W.O. (Ed.), *Manual of British Water Engineering Practice*, Heffer, Cambridge, for Inst. Water Engrs, 4th Edn, 1969.
9 Camp, T.R., Sedimentation and the design of settling tanks, *Trans. Am. Soc. Civ. Engrs*, **111**, 895–958, 1946.
10 Newitt, D.M., Richardson, J.F., Abbot, M. and Turtle, R.B., Hydraulic conveying of solids in horizontal pipes, *Trans. Inst. Chem. Engrs*, **33**, 93–110, 1955.

7

Non-uniform channel flow

Notation

Symbols will be used with the meanings given on page 111. Additional notation is as follows:

s	invert (or bed) gradient, strictly the sine of the angle of slope
x	distance measured in the direction of flow
z	height of channel bed above a datum level
c (suffix)	applied to A, T or y to signify critical depth value
n (suffix)	applied to A, m or y to signify normal depth value
E	specific energy, $y + \alpha V^2/2g$
F	specific force
T	width of water surface
$\mathscr{F}$	Froude number, $V/\sqrt{(gA/T)}$
α	Coriolis velocity distribution coefficient, for kinetic energy
β	Boussinesq velocity distribution coefficient, for momentum

7.1 Non-uniform flow in open channels

It has already been indicated, in Section 6.4, that the depth of flow in a channel or partly full pipe is rarely uniform. Ponds, changes of slope or of section, steps, etc., control depth of flow locally and, whenever the depth is different from the normal depth for uniform flow, a gradual change of depth occurs along the length of the channel. It is only by taking into account the velocity energy changes associated with the changing depth that depth of flow for a given discharge can be calculated.

Depth of flow can be related to distance measured along the channel bed by a differential equation, known as the equation of non-uniform flow, as follows.

The total energy at a cross-section of flow may be expressed (Fig. 7.1) as

$$H = z + E \tag{7.1}$$

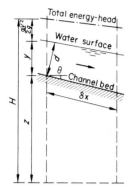

Fig. 7.1 Energy in open-channel flow

where

$$E = y + \frac{\alpha V^2}{2g}$$

Strictly, d should be used instead of y, but for the small values of θ involved in most cases the vertical depth $y = d/\cos \theta$ is a sufficiently close approximation.

Differentiating (7.1) with respect to x,

$$dH/dx = dz/dx + dE/dx$$

or $\qquad - i = - s + (dE/dy)(dy/dx)$

and $\qquad \dfrac{dy}{dx} = \dfrac{s - i}{dE/dy}$ \hfill (7.2)

The negative signs for i and s arise because slopes of the energy lines and of the channel bed are downwards in the direction of x. In a case where the bed sloped upwards, the value of s in (7.2) would be negative.

7.2 Classification of longitudinal water-surface profiles

For channel cross-sections of practical interest, y cannot be expressed directly in terms of x. Instead, equation (7.2) has to be solved arithmetically using a step method. Cases where this is warranted are not common, but it is frequently desirable to obtain a rough idea of the shape of the longitudinal surface profile. Six basic shapes occur and can be deduced from (7.2). Even when the relationship of y to x is to be calculated in detail, it is necessary to know which class of profile is involved.

In classifying surface profiles, the signs of $s - i$ and of dE/dy are important factors, as they determine the sign of dy/dx.

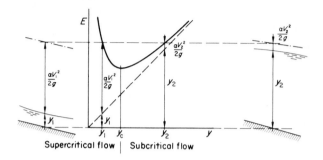

Fig. 7.2   Relationship between specific energy and depth of flow

Considering the sign of dE/dy first, E may be expressed as

$$E = y + \frac{\alpha}{2g}\left(\frac{Q}{A}\right)^2 \tag{7.3}$$

in which Q is constant and A depends on y. The relationship of y to E is shown graphically in Fig. 7.2. As an increase in y decreases the velocity energy term, two different values of y, known as alternate depths, give the same value of E except at the minimum value of E where the alternate depths coincide. The value of y which makes E minimum is known as the critical depth y_c. If the actual depth of flow (due to constraint or gradient) is greater than y_c, then the flow is termed sub-critical and if the depth is less than y_c the flow is termed super-critical, the prefixes 'sub' and 'super' relating to velocity (or Froude number, see below) rather than depth.

An expression for critical depth may be obtained by differentiating E:

$$\frac{dE}{dy} = 1 - \frac{\alpha Q^2}{gA^3} \cdot \frac{dA}{dy}$$

$$= 1 - \frac{\alpha Q^2}{g} \cdot \frac{T}{A^3}$$

in which T is the breadth of the water surface which is equal to dA/dy. For critical depth:

$$\frac{dE}{dy} = 1 - \frac{\alpha Q^2}{g} \cdot \frac{T_c}{A_c^3} = 0$$

in which A_c and T_c are the values when $y = y_c$.

Consequently

$$\frac{A_c^3}{T_c} = \frac{\alpha Q^2}{g} \tag{7.4}$$

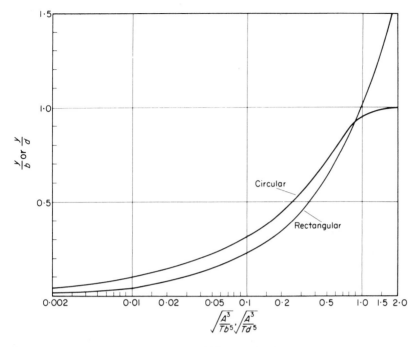

Fig. 7.3 Critical depth factors for rectangular and circular sections

and y_c may be found by seeking the value which makes A^3/T equal to $\alpha Q^2/g$. This may be done conveniently with the aid of a table or graph, Fig. 7.3, of y/b against A^3/Tb^5 in which b is some fixed dimension of the channel cross-section such as the bottom width of a rectangular or trapezoidal section or the diameter of a circular section.

It is evident from Fig. 7.2 that dE/dy is positive in sub-critical flow and negative in super-critical flow. The differential dE/dy may be written

$$\frac{dE}{dy} = 1 - \mathscr{F}^2$$

in which

$$\mathscr{F}^2 = \frac{\alpha V^2}{g(A/T)},$$

and $\mathscr{F}$ is known as the Froude number.

Flow is sub-critical for $\mathscr{F} < 1$ and super-critical for $\mathscr{F} > 1$. The Froude number is a measure of the ratio of inertial to gravitational forces, and arising from this it gives the ratio of mean velocity of flow to the celerity of small surface waves. Consequently, while waves can travel upstream in sub-critical flow, they cannot do so in super-critical flow.

Turning now to the sign of $s - i$, the normal depth of uniform flow depends on s. It is the depth at which flow would occur in a channel quite free from controls and is related to it by:

$$s = \frac{fQ^2}{2gA_n^2 m_n}$$

in which A_n and m_n are values corresponding to normal depth y_n. Methods of finding y_n have been discussed in Chapter 6. The energy gradient i depends on the actual value of y at a given cross-section of flow:

$$i = \frac{fQ^2}{2gA^2 m}$$

Consequently

$$s - i = \frac{fQ^2}{2g}\left(\frac{1}{A_n^2 m_n} - \frac{1}{A^2 m}\right)$$

Often, both A and m increase with y so that $s - i$ is positive if $y > y_n$ and negative if $y < y_n$.

Exceptions occur in sections with a closing crown, for example in circular pipes, where $A^2 m$ (or $A^2 m^{4/3}$ if Manning's formula is used, see Fig. 6.4) rises to a maximum at a certain depth and then falls again. If $A^2 m$ is greater than the running-full value $A_0^2 m_0$, there are two possible values of y for a given value of $A^2 m$. Thus if Q and s are such that $A^2 m$ falls within this range, there are two possible normal depths. They are known as conjugate normal depths. Using y_n for the lower conjugate normal depth, and y_n' for the upper value:

$$\text{for } y \quad < y_n \qquad\qquad s - i \text{ is negative}$$
$$\text{for } y_n < y < y_n' \qquad s - i \text{ is positive}$$
$$\text{and for } y_n' < y < d \qquad s - i \text{ is negative}$$

In sections to which these peculiarities do not apply it will now be evident that dy/dx is positive at depths greater than both normal and critical, since both $s - i$ and dE/dy are positive, and also at depths less than both normal and critical, since both $s - i$ and dE/dy are negative. At depths between normal and critical, $s - i$ and dE/dy are of opposite signs and dy/dx is negative, but two separate cases arises as y_n is sometimes greater, sometimes less, than y_c. This is because y_n depends on s but y_c does not. Bed slopes which make $y_n > y_c$ are termed mild, those which make $y_n < y_c$ are termed steep. The value of s which makes $y_n = y_c$ is known as the critical slope, s_c.

The shapes of profile which arise from these considerations are shown in Fig. 7.4. Where profiles approach normal depth, they approach it asymp-

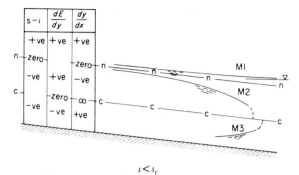

$$s < s_c$$

For $s = 0$, $y_n = \infty$ and the upper curve does not exist.
For $s < 0$ (adverse slope), y_n is fictitious. The upper curve does not exist and the middle curve has horizontal asymptote upstream.

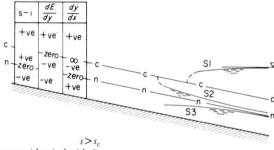

$$s > s_c$$

At $s = s_c$, M1 becomes identical with S1
and M3 becomes identical with S3

Fig. 7.4 Water-surface profiles in sections without closed crown and for $Q < Q_0$ in sections with closed crown

totically at the end of the channel remote from the control. In simple theory, profiles approach critical depth at right-angles. In channels of zero slope, normal depth is infinite and in channels of adverse slope it is entirely fictitious. Consequently, the upper mild slope profile does not occur in these and the middle one becomes asymptotic to the horizontal in the upstream direction for channels of adverse slope.

For sections with a closing crown, the same profiles occur at discharges less than the running-full discharge Q_0, but at higher discharges where there are two normal depths several additional profiles are possible, at least in theory. The additional profiles are shown in Fig. 7.5.

Where profiles in Fig. 7.5 are above the crown they represent the pressure-head line (or hydraulic gradient). The slope of the line is uniform. Strictly, since x is measured in the direction of the slope of the channel, the absolute gradient of the pressure head is i vertically to $\sqrt{(i^2 - s^2)}$ horizontally.

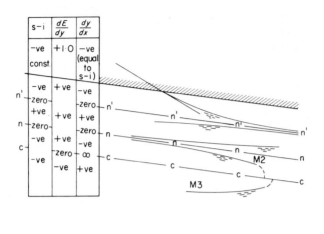

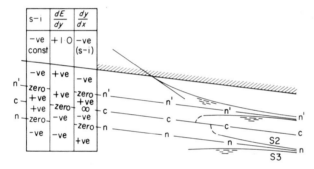

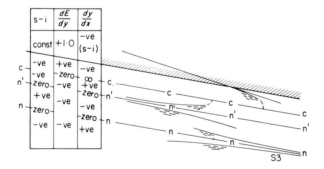

Fig. 7.5   Water-surface profiles for $Q > Q_0$ in sections with closing crown

7.3 Determining longitudinal water-surface profiles

Whichever method of obtaining the profile is to be used, it is first necessary to calculate normal depth (see p. 118), critical depth (p. 130) and the depth at the section of control. If control is exercised by a weir, sluice or constriction the depth will be determined by the head-discharge relationship of the device. Where the channel flows into or out of a pond, pond level, with due allowance for conversion of kinetic energy and for energy losses, will determine the controlled depth, and pond level may itself be determined by some hydraulic control. At a point of free discharge from the end of a mildly sloping channel such as an undrowned step or at a point where the slope becomes steep, the depth will, in simple theory, be critical. In fact it will be rather less than critical as simple theory neglects the inevitable curvature of the flow. However, it is the calculated critical depth which should be used in profile computations to avoid inconsistency.

Knowledge of normal, critical and controlled depths enables the class of profile to be identified and the curve can then be roughly sketched on a longitudinal section. For many purposes, this is sufficient. If a more accurate estimate is needed, equation (7.2) must be integrated in some way.

Since at every point along the channel i and dE/dy depend on the value of y at that point, results can be obtained directly only by choosing a series of y-values and then finding the sections (by evaluating corresponding x-values) at which the chosen values will occur. Depths cannot be calculated at specified points except by trial-and-error. While this is sometimes necessary in rivers with varying section and a procedure[1] ('indirect step') is available, methods of calculating x from y are usually adequate in sewerage and sewage-treatment design.

Programs for the computation of longitudinal surface profiles are readily available and are designed to cover wide ranges of cases and of cross-sectional shapes.

Rectangular channels are frequently used in treatment works and profiles can be computed quite quickly for these on a small calculator using the following version of the gradually-varied flow equation:

$$\Delta x = \frac{\Delta y}{f} \cdot \frac{1 - (y_c/y)^3}{\left(\dfrac{y_c}{y_n}\right)^3 \left(\dfrac{1}{2} + \dfrac{y_n}{b}\right) - \left(\dfrac{y_c}{y}\right)^3 \left(\dfrac{1}{2} + \dfrac{y}{b}\right)} \tag{7.5}$$

A series of y-values in steps of Δy is taken, covering the range of the profile. The equation is used to calculate Δx for values of y which are the mean values of successive pairs of the original series. The Δx are the computed distances between the sections where the specified depths occur. The smaller the steps, the better the approximation.

The range of y is usually small enough for the same value of f to be used throughout. In this case, only $(y_c/y)^3$ and y/b vary from step to step.

The need to compute a profile arises most often in channels of zero slope and for these the equation can be simplified to:

$$\Delta x = \frac{\Delta y}{f} \cdot \frac{1 - (y_c/y)^3}{\left(\dfrac{1}{2} + \dfrac{y}{b}\right)} \tag{7.6}$$

If the depths along the profile are large in relation to critical depth, $dE/dy \approx 1$, as can be seen in Fig. 7.2. A close approximation to the profile can then be obtained with a reduced amount of calculation by using:

$$\frac{dx}{dy} \approx \frac{1}{s - i}$$

in the form

$$\Delta x \approx \frac{\Delta y}{s - i} \tag{7.7}$$

This method is frequently used for carrying out calculations for sewage works channels but it should not be forgotten that it is acceptable only where the flow is well into the sub-critical region.

7.4 The momentum principle

For some design problems the energy principle, used above to predict variations of depth along a channel, is insufficient or inapplicable (though still valid) and the momentum principle has to be used instead or in addition.

In open channel problems the momentum principle is applied by defining a stationary control volume as shown by the broken lines in Fig. 7.6. The component of total force on the fluid in the control volume in a chosen direction may then be equated to the rate of change of momentum of the fluid within the control volume in that direction.

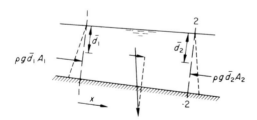

Fig. 7.6   Forces on fluid in a control volume for open channel flow

In the figure the x-direction components of the forces are as follows.

1 Forces due to the pressure of fluid where it is intersected by the boundaries of the control volume. The pressure will be hydrostatically distributed so the force may be written as the product of the pressure at the centroid of cross-section, below the surface, and the area of cross-section, A. The x-direction forces are thus

$$\rho g \bar{d_1} A_1 \quad \text{and} \quad -\rho g \bar{d_2} A_2$$

2 The x-component of the gravity force.
3 The total force due to friction at the bed and walls in the x-direction.
4 The x-components of the reactive forces due to pressure on the bed and walls e.g. at steps or changes in breadth.

As a consequence of flow through the control volume, momentum is continuously entering the volume at $1-1$ and leaving at $2-2$. The rate of momentum flow is the product of the flow of mass ρQ and the velocity, V. If it is acceptable to consider velocity to be uniform over the cross-section the momentum flow (or flux) can be written

$$\rho Q V \text{ or } \rho A V^2 \text{ or } \rho Q^2 / A$$

However, velocity will, in truth, vary across the flow and momentum flux should correctly be written $\int \rho v^2 \mathrm{d}A$, integration being applied over the cross-section. The usual practical way of dealing with this is to express the momentum flux as

$$\beta \rho Q V$$

in which V is the mean velocity Q/A or $\int v \mathrm{d}A/A$ and β, the Boussinesq velocity distribution coefficient, is

$$\int v^2 \mathrm{d}A/(V^2 A)$$

The coefficient has a value of about 1.04 for ordinary channel flows. Equating all x-direction forces to the rate of change of momentum:

$$\rho g \bar{d_1} A_1 - \rho g \bar{d_2} A_2 + \text{boundary and gravity forces}$$
$$= \beta_2 \rho Q V_2 - \beta_1 \rho Q V_1$$

Collecting terms relating to inflow and outflow sections,

$$\rho g \bar{d_1} A_1 + \beta_1 \rho Q V_1 = \rho g \bar{d_2} A_2 + \beta_2 \rho Q V_2 + \quad \text{boundary and gravity forces} \quad (7.8)$$

It is often convenient to regard the sum of the pressure force and momentum flux at a cross-section of flow as a composite characteristic of the flow at that cross-section, sometimes called 'pressure-plus-momentum'. A further step is to divide each term by ρg and call the result specific force, F thus

$$F = \bar{d} A + \frac{\beta Q^2}{gA} \tag{7.9}$$

so that

$$F_1 = F_2 + (\text{boundary and gravity forces})/\rho g$$

In some applications the boundary and gravity forces are negligible. In any event F is a useful characteristic. It has an analogous property to the specific energy, E, in that for a given flow there is a critical depth (the same as from energy considerations) which makes F a minimum, while for larger values of F, two values of depth will give the same specific force. The lower depth corresponds to a supercritical flow and the upper or sequent depth corresponds to a subcritical flow. Figure 7.7 illustrates the relationship of y with F.

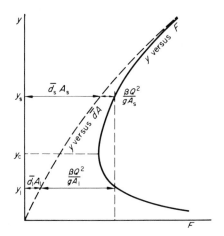

Fig. 7.7   Relationship of depth of flow with specific force

For a rectangular cross-section,

$$F = \frac{1}{2}y^2b + \frac{\beta Q^2}{gby} \tag{7.10}$$

and evaluation is straightforward.

For circular sections, graphs or tables are useful. Figure 7.8 gives the dimensionless relationship, dividing F by d^3 and using a discharge parameter $\sqrt{(\beta Q^2/gd^5)}$. For flows under pressure $y/d > 1.0$ and the relationship between y and F becomes linear since d is linear with y and A remains constant.

As with energy equations it is sometimes useful to employ the Froude number. Thus for rectangular sections

$$F = y^2b \left(\frac{1}{2} + \mathscr{F}^2 \right)$$

since

$$\mathscr{F}^2 = Q^2/(gb^2y^3)$$

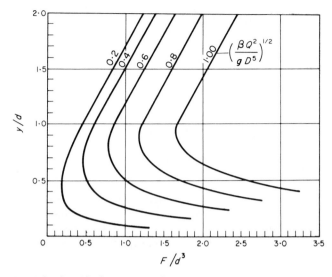

Fig. 7.8 Specific-force curves for circular sections

Dimensionlessly

$$\frac{F}{b^3} = \left(\frac{y}{b}\right)^2 \left(\frac{1}{2} + \mathscr{F}^2 \right)$$

(7.11)

7.5 Application of the momentum principle

(1) Channels receiving flow along their length

Flow leaves sedimentation tanks over a weir at the end of a rectangular tank or around the periphery of a circular tank. The flow is collected in a channel. The discharge is zero at the point most remote from the collecting channel outfall and increases linearly along the channel. In order to design such a channel, the maximum depth of flow is needed. This can be obtained as follows if bed slope is zero, if the width is uniform and if friction can be assumed negligible, see Fig. 7.9.

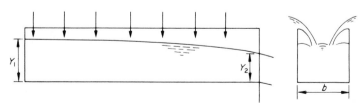

Fig. 7.9 Channel with spatially increasing flow

Considering the upstream and downstream ends of the channel,

$$F_1 = F_2 \tag{7.12}$$

There is no gravity force in the x-direction because the bed is horizontal. There are no direct x-direction forces from the walls because the channel is prismatic and friction is being neglected.

From (7.11) and (7.12)

$$\left(\frac{y_1}{b}\right)^2 \left(\frac{1}{2} + \mathscr{F}_1{}^2\right) = \left(\frac{y_2}{b}\right)^2 \left(\frac{1}{2} + \mathscr{F}_2{}^2\right)$$

but $\mathscr{F}_1{}^2 = 0$ since there is no flow at the upstream end.

Thus $y_1 = y_2\sqrt{(1 + 2\,\mathscr{F}_2{}^2)}$ \tag{7.13}

Conditions downstream of the outlet will control y_2 and $\mathscr{F}_2{}^2$. These will be known from the design of the downstream channel. Frequently the outlet is in the form of a step so that y_2 is the critical depth and provided the step is not drowned, downstream conditions will not have any influence. In this case $\mathscr{F}_2{}^2 = \mathscr{F}_c{}^2 = 1.0$
and so

$$y_1 = y_c\sqrt{3}$$

Friction losses can be taken into account by using the following equation of Wen-Hsuing Li:[2]

$$y_1 = y_2(1 + a\,\mathscr{F}_2{}^2)^b \tag{7.14}$$

in which

$$a = \frac{8 + 3\alpha}{4} \quad \text{and} \quad b = \frac{4 + \alpha}{8 + 3\alpha}$$

$$\alpha = h_f/(\mathscr{F}_2{}^2 y_2) \quad \text{or} \quad fL/(2m_2)$$

and h_f is the friction loss in a channel of depth y_2, length L and hydraulic mean depth m_2. It will be noted that for zero friction, α becomes zero and equation (7.14) becomes identical with equation (7.13).

The formulae given above apply to level channels. For sloping channels, a formula may be derived by assuming that the water surface is parabolic longitudinally:

$$\frac{y_1}{y_2} = ((1 - \tfrac{1}{3}G)^2 + 2\mathscr{F}_2{}^2)^{1/2} - \tfrac{2}{3}G \tag{7.15}$$

in which G is sL/y_2. This formula gives results very close to those obtained by Li from detailed computation. Friction is neglected but can be estimated by comparing results from equations (7.13) and (7.14).

The fundamental equation for spatially increasing flow is

$$\frac{dy}{dx} = \frac{s - i - 2Qq'/(gA^2)}{1 - \mathscr{F}_2} \tag{7.16}$$

in which $q' = dQ/dx$ and Q, A and $\mathscr{F}_2$ all vary with x.

(2) Channel junctions

In carrying out design calculations for sewage works channels, the depth of
flow at the downstream end of a junction will usually have been determined
in the calculations for the downstream channel. The purpose of applying the
momentum principle to the junction is then to find the depths in the
approach channels so that design can proceed upstream.

A form of equation suited to this purpose will be described initially in
relation to Fig. 7.10a. The cross-section of the approach channels will be
among the items the designer needs to fix and it is suggested that this will be
done mainly with regard to the mean velocities. Quite often the velocities will
be made equal to that downstream, V_2, but to be more general let us call the
velocities in the approach channels $k_1 V_2$ and $k_j V_2$.

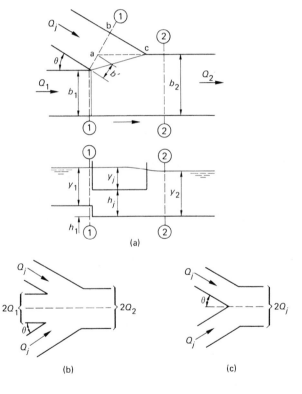

(a)

(b)   (c)

Fig. 7.10   Channel junctions

With certain assumptions relating to forces at cross-section 1–1, detailed
below, the x-direction force momentum equation can be written:

$$\tfrac{1}{2}\rho g(y_1 + h)^2 b_2 + \rho Q_1 k_1 V_2 + \rho Q_j k_j V_2 \cos\theta = \tfrac{1}{2}\rho g y_2^2 b_2 + \rho Q V_2$$

The assumptions all relate to the first term and are as follows:

1 That the water level at 1–1 is common to both branches so that

$$y_1 + h_1 = y_j + h_j$$

2 That the reactive forces from the steps upon the water are the same as they would be if pressure distribution were hydrostatic.

3 That the x-direction force components on the vertical sections ab and bc are negligibly different from each other and thus balance.

With these assumptions, the force terms for section 1–1

are $\frac{1}{2}\rho g(y_1 + h)b_1$ for the main channel

and $\frac{1}{2}\rho g(y_j + h_j)b' \cos \theta$ for the branch.

Since

$$b_1 + b' \cos \theta = b_2 \qquad \text{and} \qquad y_1 + h = y_j + h_j$$

the sum of these terms is

$$\frac{1}{2}\rho g(y_1 + h)^2 b_2$$

Dividing the equation by $\frac{1}{2} g y_2^2 b_2$, substituting the downstream Froude number and re-arranging:

$$\left(\frac{y_1 + h}{y_2}\right)^2 = 1 + 2 \, \mathscr{F}_2^2 \left(1 - \frac{Q_1}{Q_2} - k_j \, \frac{Q_j}{Q_2} \cos \theta\right) \qquad (7.17)$$

This equation can be applied to a double branch junction as in Fig. 7.10b by regarding Q_1 and Q_2 as *halves* of the main channel flows. For a Y-junction, Fig. 7.10c, Q_1 is zero so

$$\left(\frac{y_1 + h}{y_2}\right)^2 = 1 + 2 \, \mathscr{F}_2^2 (1 - k_j \cos \theta) \qquad (7.18)$$

It is interesting to note that with $\theta = 90°$ the result is similar to that for spatially increasing flow:

$$\left(\frac{y + h}{y_2}\right)^2 = 1 + 2\mathscr{F}_c^2 \qquad (7.19)$$

However, though experimental work has confirmed that the Y-junction equation is reasonably accurate, the 90° version underestimates upstream depth. Eddying flow just downstream of the corners, leaving a contracted flow in the middle, becomes quite pronounced when θ becomes a right angle.

A possible design strategy using the above equations is to begin by assigning values to k_1 and k_j. From these, values of y_1, y_j, b_1 and b_j can be obtained. The appropriate equation can then be used to find $(y_1 + h_1)/y_2$ and hence the height of the step. A few trials may be needed to obtain satisfactory proportions. Often, making all velocities the same ($k_1 = k_j = 1$) and also the y/b ratios will result in a satisfactory design, velocities and y/b ratios having economic implications as discussed in Section 14.4. An example is given in Section 14.5.

(3) The hydraulic jump

Super-critical flows are associated with controls at the upstream end of a channel, while sub-critical flows are associated with controls at the downstream end. Consequently, it is possible to have a length of channel with upstream control tending to cause super-critical flow and down-stream control tending to cause sub-critical flow. Such a case is shown in Fig. 7.11. Due to the change from mild to steep slope at C, the depth of flow at that section is critical and, as BC is of mild slope, the flow upstream of C is sub-critical. However, the steep reach above A is delivering super-critical flow into BC. In the case shown there is an abrupt change from super-critical to sub-critical flow, known as a hydraulic jump, at a point between B and C.

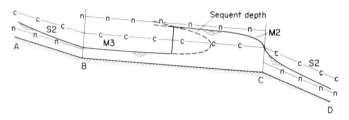

Fig. 7.11 Hydraulic jump

The jump would not necessarily occur on BC. For example, if the gradient of AB were less steep, the jump could form on AB, sub-critical flow occurring for a short distance upstream of B. On the other hand, if AB were more steep, super-critical flow could persist along the whole of BC without the formation of a jump.

The occurrence of a hydraulic jump may be predicted by using a control volume which intersects the flows just upstream and just downstream of the jump, see Fig. 7.12. The control volume is usually considered to be small

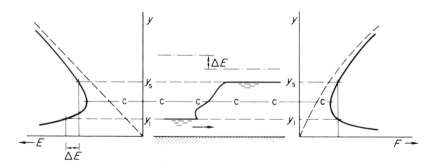

Fig. 7.12 The hydraulic jump — force and energy relationships

enough for friction and gravity forces to be neglected. Specific forces upstream and downstream can then be equated:

$$F_1 = F_2$$

A great deal of energy is dissipated in the jump. This is the reason why the energy principle cannot be used to relate the upstream and downstream depths. However, if the energy principle is applied in addition to the momentum principle the loss of specific energy, E, can be predicted. This is shown graphically in Fig. 7.12. The specific force shown on the right-hand side relates the upstream depth y_1 to the downstream or sequent depth y_s, both having the same specific force. The sequent depth is always less than the upper alternate depth (which has the same specific energy as y_1) and consequently, since it is sub-critical, has a smaller specific energy than the upper alternate depth. This may be demonstrated as shown in Fig. 7.12 by transferring y_s across to the specific energy graph on the left and comparing its specific energy with that of y_1. The mechanical fluid energy dissipated causes a rise in the thermal energy of the flow generated by the internal friction of the eddies giving a rise in temperature. In addition there are fluctuating pressures on the channel bed beneath the jump which can be damaging if it is not massively constructed.

If a graph or table is available relating depths to specific force for the channel section in question the theoretical position of the jump is easily found as shown in Fig. 7.11 by choosing a number of points on the super-critical profile (an M3) and using the specific force graph to find the sequent depth at each point. The curve of sequent depth is plotted on the scale drawing of the profiles. In theory the jump occurs where the sequent depth curve intersects the sub-critical profile (an M2).

If specific force curves are not available the method shown in Fig. 7.13 may be used. Here values of F calculated from depths at chosen positions along the profiles are plotted above the profiles to some suitable scale. The

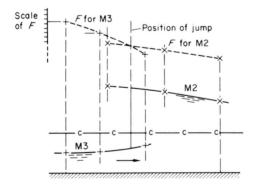

Fig. 7.13 Theoretical position of hydraulic jump determined by plotting F-values for super- and sub-critical streams on a longitudinal section

intersection of the *F* curves gives the position of the jump. Both methods can of course be used arithmetically instead of graphically if desired.

The intersections of the curves determining jump position in the figures have been made fairly well-conditioned for purposes of clarity but in practical cases they are often ill-conditioned so no great precision is obtainable. Quite small changes in the estimated depths of the profiles will cause large shifts in the position of the jump as will small changes of discharge or controls in an actual channel.

Though in simplified theory the jump is vertical, in truth it is in the form of a roller and extends over a length of channel. The length has been found to relate to the Froude number of the approach flow and empirical data are available for rectangular section channels.[1] The form of the jump, whether weak, strong or undular, is also related to the approach flow Froude number.

If it is desired to localise the jump over a range of flows a sill or dentations on the bed of the channel, or a local widening to form a stilling basin may be employed. In irrigation practice energy dissipators of this type have been well developed and can be used in sewerage schemes for cases where steep surface water outfall sewers have to discharge to open water courses. Specialised books on open channel flow should be consulted for details.

In a roofed conduit a super-critical free-surface flow can jump to a running-full condition as shown in Fig. 7.14.

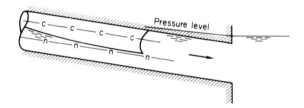

Fig. 7.14 Hydraulic jump to full-bore flow

The formulation of the hydraulic jump in a rectangular section open channel is particularly simple and is sometimes useful in treatment works design.

Using $F_1 = F_2$ for a rectangular section,

$$\frac{1}{2} by_1^2 + \frac{Q^2}{gby_1} = \frac{1}{2} by_2^2 + \frac{Q^2}{gby_2}$$

Dividing by $\frac{1}{2}by_1^2$ and substituting $\mathscr{F}_1^2 = Q^2/(gb^2y_1^3)$ and $r = y_2/y_1$,

$$1 + 2\mathscr{F}_1^2 = r^2 + \frac{2\mathscr{F}_1^2}{r}$$

from which:

$$r = \tfrac{1}{2}\{\sqrt{(1 + 8\mathscr{F}_1^2)} - 1\}$$

7.6 Side weirs

Where a channel has a weir in one or both side walls, flow spills from the channel, the rate of flow thus decreasing along the main channel. It would be incorrect to assume that specific force remained constant along the main channel as was done for spatially increasing flow since the spilled flow retains some x-direction momentum. Indeed if specific force is taken to be constant, the depths of flow obtained imply a gain in energy along the flow. The use of the momentum principle being precluded, it is assumed that specific energy remains constant apart from ordinary friction losses, and an equation of spatially decreasing flow can be derived:

$$\frac{dy}{dx} = \frac{s - i - Qq'/(gA^2)}{1 - \mathscr{F}^2} \tag{7.20}$$

Strictly, q', the rate of change of flow along the channel, is negative because the flow is decreasing. The third term of the numerator is thus an addition. With a positive sign for this term as in some versions, the algebraic sign of q' is ignored.

Several modes of flow can be deduced from this equation. The one most common in sewerage and sewage schemes is shown in Fig. 7.15. Spill from the channel results in a drawdown profile upstream of the weir but it is an essential feature of the mode of flow shown that flow is sub-critical throughout, the weir sill being substantially above critical depth. The increase in depth along the weir follows from the conservation of energy, flow being sub-critical.

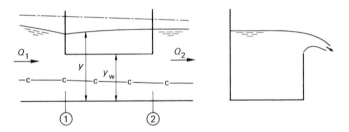

Fig. 7.15 Flow over a side weir

Given the dimensions of the channel and of the weir together with the upstream flow it is a difficult task to calculate how much water remains in the main channel and how much spills. However, at sewage treatment works the flow able to proceed is almost invariably limited by a penstock automatically controlled by a gauging flume and the problem is to find how long to make the weir so that upstream depths are not too large.

The length of storm weir at a treatment works can be found by carrying out a step-by-step calculation with the following equations:

$$\Delta y = \frac{\Delta Q}{Q/y - g(by)^2/Q} \tag{7.21a}$$

$$\Delta x = \frac{\Delta Q}{K(y - y_w)^{3/2}} \tag{7.21b}$$

The first of these is equation (7.20) applied to a rectangular channel with zero bed slope and negligible friction, re-arranged to relate a change of depth, Δy, along a portion of weir length, Δx, over which a flow, ΔQ, spills.

The second equation is a weir flow equation re-arranged to enable crest length, Δx, to be calculated from the discharge, ΔQ, the weir coefficient, K, and the head $(y - y_w)$.

The method of use is to divide the total flow to be spilled into a number of equal parts, ΔQ, from which a series of mean flows $Q_1 - \frac{1}{2}\Delta Q$, $Q_1 - 1\frac{1}{2}\Delta Q$, etc. can be obtained. Equation (7.21a) is then applied to the first of these with y as the intended channel depth near the end of the weir. The calculated Δy enables the depth to be adjusted for the next portion of the weir, for which the second flow is used, and so on. Having thus obtained a series of mean depths corresponding to successive sections of the weir, each discharging ΔQ, equation (7.21b) is used to obtain the Δx, the sum of which gives the necessary length of the weir.

The weir coefficient may be obtained from the following expression deduced by Ackers from the experimental results of Fraser:

$$C = 0.78(1 - 0.44\mathscr{F})$$

in which $\mathscr{F}$ is the Froude number of the flow at the upstream end of the weir. The coefficient K is related to a dimensionless coefficient, C by

$$K = \tfrac{2}{3}C\sqrt{2g}$$

The change of depth along a storm weir is typically quite small and has frequently been taken to be negligible, length being calculated from

$$Q = KLh^{3/2}$$

with h as the intended head over the weir. Before the Ackers formula was available K would usually have been given the broad-crested weir value of 1.705 m$^{1/2}$/s.

Downstream flow in a number of forms of sewer storm overflows is controlled by an orifice or by a short length of pipe. Though there is no automatic control, water level for the expected maximum flow can be estimated from the application of the methods above, though the cross-section of flow is seldom rectangular. The section between the dip plates may be a good enough approximation.

In low side weir-overflows, the weir sill is below critical depth and a different mode of flow from that shown in Fig. 7.15 applies.

7.7 The critical depth flume for flow measurement

Standard proportions[3] for this device are shown in Fig. 7.16. It is recommended that to avoid risk of blockage the throat breadth b, should be at least 100 mm. Approach flow should be sub-critical with $\mathscr{F}$ preferably not greater than 0.5. At greater values, surface waves tend to occur. An upper limit for the upstream depth is 2 m.

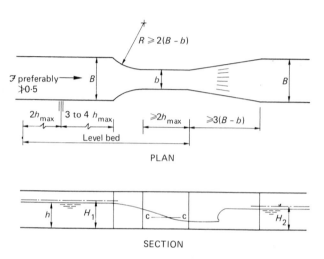

Fig. 7.16   Critical depth flume for flow measurement

As flow becomes constricted towards the throat, velocity and kinetic energy increase resulting in a decrease in potential energy, that is in depth. Provided depth can pass through critical depth at the throat, with a hydraulic jump back to sub-critical flow in the divergence, upstream depth is independent of downstream depth.

In a rectangular channel

$$y_c^3 = Q^2/(gb^2)$$

and

$$\frac{V_c^2}{2g} = \frac{Q^2}{2gb^2y_c^2} = \frac{y_c}{2}$$

Thus the specific energy for critical depth is $1.5y_c$.

Neglecting losses we may write

$$H_1 = \frac{3}{2}y_c = \frac{3}{2}\left(\frac{Q^2}{gb^2}\right)^{1/3}$$

giving $Q = (\frac{2}{3})^{3/2}g^{1/2}bH_1^{3/2} = 1.705\,bH_1^{3/2}$ in metre second units.

Friction will reduce the coefficient a little. Note that H_1 is the upstream specific energy:

$$H_1 = h + V^2/(2g)$$

However $V^2/(2g)$ is usually quite small.

In order to ensure that a jump forms in the divergence, the ratio H_1/H_2 must not exceed the value shown below according to the length of the divergence.

Length of divergence	$3(B - b)$	$6(B - b)$	$10(B - b)$	$20(B - b)$
H_1/H_2 not greater than	0.74	0.80	0.83	0.91

7.8 The vortex drop

This device is shown in Fig. 7.17. Flow enters along the approach channel causing swirling, free vortex, motion in the volute-shaped chamber. Super-

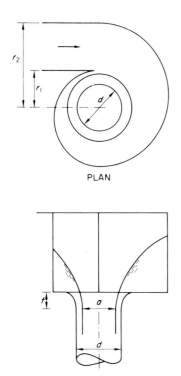

PLAN

SECTION

Fig. 7.17 The vortex drop

imposed on the vortex motion is a radial flow towards the drop pipe. Over a very wide range of conditions, due to the swirl, water descends the drop pipe in a helical path, clinging to the walls and leaving a core of air in the middle.

If the outlet to the drop pipe is restricted it may run full near the outlet and there will be a transition from air-core flow to full-bore flow in the form of an annular hydraulic jump. Provided that air-core flow persists below the throat section, discharge is independent of downstream conditions and the approach channel depth for a given discharge can be reliably predicted. The relationship between depth and discharge is almost linear, except for very small discharges at which radial weir flow occurs at the lip of the drop.

Vortex drops were installed at Portland, Oregon, as a means of controlling depths of flow alongside storm overflow weirs.[4] They have been used more recently in Britain for drop manholes[5] and have potential use as gauging structures and, in the rare instance where enough head is available, as depth controls for grit channels.

It is advisable for the chamber to have a volute shape, though it does no harm to use circular arcs meeting tangentially to approximate to this shape, Fig. 7.18. Experiments[6] with a circular chamber centred on the drop pipe revealed a more complex flow situation which defied analysis and gave results which departed markedly from those given by the theory below. Also, over certain ranges of flow, a large radial wave rotated in the chamber and set up a train of waves in the approach channel.

A theoretical treatment evolved by Ackers and Crump[7] enables the specific energy and hence the depth of flow in the approach channel to be related to the discharge, for chosen dimensions. Though the treatment

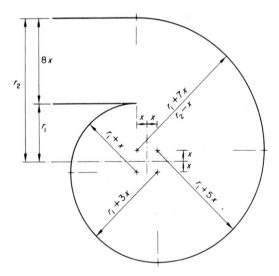

Fig. 7.18   Use of circular arcs to form vortex chamber

neglects energy losses, the theoretical relationship gives a very close approximation to experimental observations.

The dimensions which are material, r_1, r_2, d and t are shown on Fig. 7.17.

The specific energy E cannot be related to the discharge Q directly but only through a set of simultaneous equations which involve two additional variables. These are the circulation in the vortex chamber:

$$C = vr$$

where v is the tangential component of velocity at radius r; and the fractional air-core

$$f = \frac{a^2}{d^2}$$

where a is the diameter of the air-core.

There are three simultaneous equations.

1 Relating E, Q and C by considering inlet conditions.
2 Relating Q, C and f by considering conditions at the throat.
3 Relating E, C and f, using the condition of maximum discharge for given energy.

The three equations are combined to eliminate Q and C so that E is related to f by

$$\frac{E}{d}(G_1 - F_1G_2) = F_2 + \frac{t}{d}F_1G_2 \tag{7.22}$$

in which

$$G_1 = \log_e (r_2/r_1)$$
$$G_2 = \frac{d^2}{r_2^2}\left(\frac{r_2^2}{r_1^2} - 1\right)$$

and

$$F_1 = f^2/4(1 + f)$$
$$F_2 = \frac{\pi}{4}\left(\frac{1}{f} - 1\right)\sqrt{2(1 - f)}$$

The second and third equations are combined to eliminate C so that Q, E and f are related by

$$\frac{Q}{d^{5/2}\sqrt{g}} = F_2\sqrt{\left\{4F_1 \cdot \frac{E + t}{d}\right\}} \tag{7.23}$$

To produce an equation relating Q and E directly it would be necessary to eliminate f from (7.22) and (7.23) but this is not possible. Instead, a series of f-values is taken. These are used firstly to find corresponding E/d values from equations (7.22) and then, together with the E/d values, to find corresponding values of $Q/(d^{5/2}\sqrt{g})$ from equation (7.23).

The expressions for F_1 and F_2 are independent of the dimensions of the chamber and the same series of values can be used in all cases. Values are given in Table 7.1.

Table 7.1   Vortex drop functions

$$F_1 = \frac{f^2}{4(1+f)}$$

$$F_2 = \frac{\pi}{4}\left(\frac{1}{f} - 1\right)\sqrt{2(1-f)}$$

f	F_1	F_2
(1.00)	(0.125)	(0.00)
0.95	0.116	0.013
0.90	0.107	0.039
0.85	0.098	0.076
0.80	0.089	0.124
0.75	0.080	0.185
0.70	0.072	0.261
0.65	0.064	0.354
0.60	0.056	0.468
0.55	0.049	0.610
0.50	0.042	0.785
0.45	0.035	1.002
0.40	0.029	1.29
(0.00)	(0.000)	(∞)

In designing a vortex drop, the requirements which have to be met are usually a given value of approach channel breadth $r_2 - r_1$ and a given pair of values for E and Q. The dimensions which have to be found are then r_2, d and t. The equations are too complex for these dimensions to be considered singly and there is no alternative to design by trial. Since the drop pipe must be of a commercially available diameter, a trial value of d is the starting-point. Constructional factors impose some restriction on t (a bell-mouth fitting may be used) so this is fixed according to the value of d. For a trial value of r_2, enough of the E–Q relationship is worked out to enable the value of E at the design flow to be estimated. Only a few points need be calculated as the relationship is nearly linear. If the value of E is unsatisfactory a new value of r_2 is taken.

To enable an economic choice to be made, suitable values of r_2 may have to be sought for a selection of values d.

Twin vortex drops (Fig. 7.19) have some attractions. Each drop takes half the flow and r_2 is measured to the centre of the approach channel.

For purposes of making comparative rough designs to find a feasible solution, two points on the E versus Q graph can be very quickly calculated as follows:

For $f = 0.5$,

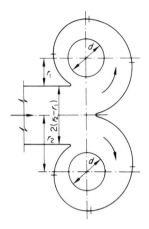

Fig. 7.19 Twin vortex-drop chamber

$$\frac{E}{d} = \frac{(\pi/4G_1) + (t/d)}{1 - (G_2/24G_1)} - \frac{t}{d}$$

and $\quad \dfrac{Q}{\sqrt{(gd^5)}} = \dfrac{\pi}{4\sqrt{6}} \sqrt{\left(\dfrac{E + t}{d}\right)}$ $\qquad\qquad$ (7.24)

A second point which is fictitious but which establishes the slope of the roughly linear relationship is obtained by putting $f = 1.0$. In this case Q is zero and

$$\frac{E}{d} = \frac{t}{d} \cdot \frac{1}{(8\ G_1/G_2 - 1)} \qquad\qquad (7.25)$$

It is also useful to note that (E/d) is asymptotic to (F_2/G_1) as f becomes smaller or E and Q become larger.

7.9 Steeply sloping sewers

In hilly districts sewers laid at slopes near to those of the ground surface inevitably involve high-velocity flows which require special consideration.

One possible consequence of high velocities is the erosion of pipe material by gritty matter in the flow. Until a few years ago many designers considered that there was an upper safe limit to the design velocity. Opinions as to the limit ranged from about 3 to about 5 m/s. Thus for each diameter of pipe there was an upper limit to the gradient, the limit becoming less as diameter increased. Where the ground sloped at a steeper gradient, the slope was negotiated in a step-wise manner using back-drop manholes, Fig. 4.4.

Alternatively, an erosion-resistant material such as cast-iron was used for the pipes.

Research commissioned by the Construction Industries Research and Information Association[8] revealed that the amount of erosion depended on the volume of grit-laden flow passing along the pipe in a given time and was independent of the velocity of flow. However, the velocity did influence the location of the erosion. At high velocities erosion was spread fairly uniformly over the wetted perimeter of flow while at low velocities a similar amount of erosion (in terms of mass of material removed) was concentrated near the invert. Thus the extent to which erosion may be a hazard appears to be more serious at low velocities than at high ones. In general it appeared from this work that straight pipelines of the usual sewer materials (clayware and concrete) are unlikely to suffer appreciable erosion. Velocities up to 10 m/s are now considered to be acceptable.

Erosion of the pipe itself may not be the only hazard, however. At steep gradients the flow will have high kinetic energy which will have to be dissipated at the foot of the slope. The flow will be super-critical if the slope is steep enough for normal depth to be below critical depth. If a steep slope is followed by a mild slope (in the hydraulic sense) a hydraulic jump will occur either near the end of the steep slope or in the mild slope if it is long enough. The jump will move back and forth in response to changes of flow unless special steps are taken (see page 145) to localise it. A hydraulic jump leads to some erosion hazard and also to pounding of the channel floor beneath the jump but whether the effects are likely to be serious in a typical sewer is not known. The power dissipated in a hydraulic jump is $\rho g Q \Delta E$ in which ΔE is the specific energy loss and is related to the Froude number of the approach flow. It would appear to be prudent to limit the approach flow Froude number (which will limit the gradient of the steep pipe) unless particularly durable and massive construction is to be used. It is not easy to say what the limit should be but it may be noted that for a rectangular channel, ΔE exceeds 26% of the approach flow specific energy when the approach flow Froude number exceeds 3.0.

Cavitation is a cause of erosion at high velocities so sharp corners in the bed or walls should always be avoided.

The problems with steep sewers are not confined to the foot of the slope. Where slope changes from hydraulically mild to hydraulically steep, the diameter should not be reduced without considering the kinetic energy requirements of the flow in the steep pipe. Figure 7.20 shows what can happen if advantage of the steeper slope is taken to reduce the pipe diameter. The upstream pipe is forced to surcharge to provide the higher specific energy needed by the flow in the small steep pipe. A possible remedy is to lower the top end of the steep pipe, contriving a smooth drop at the manhole (Crump[9] gives a good example of such a design). In most cases this will flatten the gradient to an extent depending on the length of the steep section so a larger pipe may have to be used. The other remedy is simply to use a larger pipe still, without a drop. Usually the pipe will need to be nearly

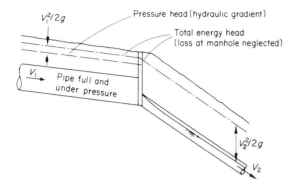

Fig. 7.20 Steep, small diameter pipe causing surcharge in upstream pipe which is larger and has a flatter slope. The free-surface flow in the steep pipe is likely to alternate with a just-full or slightly overfull flow with intermittent air gulping

as large as that of mild slope. To keep the diameters the same for both slopes is always safe and indeed the rule of thumb at one time was never to reduce diameter in the direction of flow.

The flow at bends in super-critical flow sewers is of some interest. The governing criterion is V^2/gR in which V is the mean velocity of flow and R is the radius of curvature of the bend. When this number is less than 1.0, super-elevation in the form of a train of standing waves occurs and these persist beyond the bend becoming gradually damped out. When $V^2/(gR) > 1.0$, the flow makes a complete rotation of the pipe cross-section, clinging to the pipe wall and leaving an air core in the middle. In both cases the flow remains super-critical so the presence of the bend has no influence on upstream water levels. Consequently there is no need to use a larger diameter for a given flow as compared with a straight pipe. However, it may be wise not to have an open manhole channel at the bend if $V^2/(gR)$ is greater than 1.0.[10]

7.10 Hydraulic models

In some features of sewerage and sewage-treatment schemes, flow is too complex for performance to be predicted by calculation. Such features include drops, ramps, large junctions, convergences, divergences and tail-bays of outfalls. Scale models can be usefully employed to check or to improve the design. A model which is a smaller-scale replica of the proposed design is used. It is rarely necessary to use a vertical scale which is different from the horizontal scale for sewerage structures.

In order to replicate depths of flow on the model corresponding to design discharges on the prototype it is necessary to ensure that the forces causing and resisting flow are in the same ratio for both model and prototype. When the flow has a free surface, as in the various circumstances covered in this chapter, the ratio of the inertial forces to the gravitational forces is of primary importance and this is given by the Froude number. Viscous and surface-tension forces will play their part and the ratios of these to inertial forces are given by Reynolds and Weber numbers respectively. Strictly, model tests should be carried out in such a way that all these numbers, and any others which are significant, are the same for both model and prototype. This would be possible only by adopting impracticable measures such as using a fluid less viscous than water, altering the force of gravity, working with atmospheric pressure different from normal, etc. However, provided the scale is not too small, the effects of viscosity and of surface tension are usually negligible in sewerage devices and the Froude number can be used as the sole criterion.

Using the Froude number, discharge rate is chosen for the model so that:

$$\frac{V_m}{\sqrt{(L_m g)}} = \frac{V_p}{\sqrt{(L_p g)}}$$

in which L_m is some length on the model corresponding to L_p on the prototype and V_m and V_p are velocities at corresponding points in the model and prototype.

Thus

$$\frac{V_p}{V_m} = \sqrt{\frac{L_p}{L_m}} = \sqrt{r}$$

in which r is the scale ratio.

Further, since

$$\frac{Q_p}{Q_m} = \frac{A_p V_p}{A_m V_p} = \frac{L_p^2 V_p}{L_m^2 V_m}$$

the discharge ratio is given by

$$\frac{Q_p}{Q_m} = r^{2.5}$$

Strictly the frictional characteristics of the model should be scaled down from those of the prototype. This is not usually practicable in a quantitative fashion but materials used for models are normally smoother than those used for actual structures. The features of sewerage schemes mentioned at the beginning of this section are all instances where large changes of velocity occur over relatively short distances and frictional effects are of very low significance.

Nevertheless, the neglect of friction, of viscosity and of surface tension means that there will be discrepancies of behaviour between model and prototype. These are known collectively as scale effects. In model work where they are significant, such effects can be assessed by testing models of different scales and extrapolating the results.

Since sewerage structures are fairly small, quite large scales can often be used for models (of the order of 4:1 to 12:1) and investigation of scale effects is seldom warranted in view of the uncertainties in estimating flows, etc., which will in any case have to be covered by a safety margin. Care should, however, be taken that the scale is not so small that flow is laminar in the model or that weir flow at low heads is within the range where surface tension is significant.

It should be clearly understood that models are of no use, other than by extrapolation of results obtained at various scales, for investigating roughness values or discharge coefficients. They are of dubious value where air entrainment is involved, for example, in siphons.

In models of short hydraulic structures, the depth of the flow delivered to the model or the depth of the flow leaving the model may have an influence on the flow in the structure. Since the channels or pipes leading to and from the model may have different frictional characteristics and lengths from those of the prototype, it is necessary to adjust their gradients until the depth of flow calculated for the prototype is reproduced. As friction in the prototype is a matter of guesswork, it may be advisable to test the model under conditions representing different estimates of depth.

It is especially important to model approach depth when the entering flow is super-critical and to model outlet depth where downstream conditions impose sub-critical flow.

Quite apart from the need to adjust inlet and outlet conditions, the model itself should be designed so that modifications can be easily made. It is often worthwhile to carry out tests on the separate hydraulic components of a structure and develop the design in stages so that modification of the complete model is minimised and alternatives are fully explored.

Where it is desired to view the flow in the model, perspex is a useful material. The available diameter of perspex tube will dictate the precise choice of scale where transparent pipes are needed. Elsewhere, sheet metal (brass, aluminium, steel) or varnished marine-grade plywood can be used. Complicated curved shapes are best made from hardwood by pattern-making techniques. Paraffin wax may also be used for curved work and is easy to modify. Modelling clay is unsuitable as it dries out and cracks when the model is not in use.

Structures which are easy to build in concrete and brickwork are surprisingly difficult to model in perspex, metal or timber especially if subsequent modification is to be provided for. Occasionally, it may be worthwhile to design the structure so that, while it can still be cheaply built of concrete, model making is not too expensive.

Since sewerage structures are relatively small, the cost of model investigations may be more than a small fraction of the cost of the structures. Alternatives to model investigations are generous design to cover the margin of uncertainty in calculation or the provision of adjustment on the prototype. The costs of these expedients need to be carefully weighed against the cost of a model. It should not be forgotten that, though the structure to be modelled may be relatively small, its performance may affect the design of much more extensive parts of the scheme which do not themselves need model investigations.

Though there may occasionally be a case for using a model merely to obtain independent confirmation of the results of calculation, there is seldom any need to use a model where experimentally confirmed theory covering the range involved is available. Calculation is, of course, quite cheap and should be used as far as possible to compare alternatives and to obtain a clear idea of the uncertainties which need experimental investigation both before and during the model tests.

References

1   Chow, Ven Te, *Open-Channel Hydraulics*, McGraw-Hill, 1959.
2   Li, Wen Hsuing, Open channels with non-uniform flow, *Trans. Am. Soc. Civ. Engrs*, **120**, 255–74, 1955.
3   British Standards Institution, BS 3680, Part 4C, BSI, London, 1973.
4   Stevens, J.C. and Kolf, R.C., Vortex flow through horizontal orifices, *Trans. Am. Soc. Civ. Engrs*, **124**, 888, 1959.
5   Hale, H.T. and Dyer, E.A., The subsidence of Fylde Street, Farnworth, *Proc. Inst. Civ. Engrs*, **24**, 207–222, Feb. 1963.
6   Yalaju-Amaye, S.D., *Investigations of Vortex Flow over Circular Weirs and Related Devices*, M.Sc. Tech. thesis, University of Manchester, 1963.
7   Ackers, P. and Crump, E.S., The Vortex Drop, *Proc. Inst. Civ. Engrs*, **16**, 433–442, Aug. 1960.
8   Vickers, J.A., Francis, J.R.D. and Grant, A.W., Erosion of Sewers and Drains, *C.I.R.I.A. Research Report* No. 14.
9   Crump, E.S., Points of interest in the design of a steeply graded pipeline, *Proc. Inst. Civ. Engrs*, Pt. III, 3, 861–889, Dec. 1954.
10 Rashid, A.L., *Effects of curvature upon super-critical open-channel flow in conduits of circular section*, Ph.D. thesis, University of Manchester, 1975.

8

Outline of sewage treatment

8.1 The need for sewage treatment

Decaying organic matter is obnoxious and harbours disease. In modern civilisation the method adopted for removing organic waste, other than dry matter or refuse, is the water-carriage system. The provision of sewers enables water polluted in other ways to be collected for treatment as well.

Having diluted waste matter with water for purposes of unobtrusive transport, the liquid has to be disposed of. Though, in a few cases, a good deal of it can be lost by evaporation or by percolation into the ground, in most civilised countries it can be got rid of only by allowing it to flow into a river or into the sea.

The effect of discharging sewage into all except the largest rivers would be merely to transfer the objectionable aspects of decay to the river and there-fore some form of treatment must be undertaken to render the polluted water more suitable for discharge. The form of treatment normally adopted consists of providing an environment in which natural processes of decay can be intensified and controlled so as to take place in the least objectionable manner.

8.2 The cycle of growth and decay

In order to appreciate the objects and methods of sewage treatment more fully, it is necessary to understand the cyclical nature of growth and decay. This is shown diagrammatically in Fig. 8.1. Living things need energy to support life and for growth. Green plants obtain energy from sunlight and use the energy to synthesise carbohydrates from water and carbon dioxide, needing small quantities of other elements in the process. Animals obtain their energy by using plants or other animals as food. Where food is used as a source of energy, the reaction is one of oxidation in which atmospheric oxygen is used to 'burn' organic material, leaving behind a residue of carbon dioxide, water and other oxidation products. These are exhaled or excreted together with those parts of the food intake which have no nutritional value to the organism concerned.

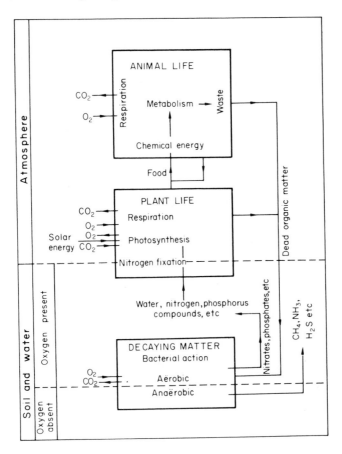

Fig. 8.1 The cycle of growth and decay

Products of excretion and other dead organic matter can be used as food by simple organisms such as bacteria, and the changes which they effect in the processes of oxidation enable certain important constituents to be reused in the cycle. The process of decay is therefore a necessary link in the endless chain. The circulation of carbon and nitrogen compounds is an important aspect of the cycle.

Decay can occur in two different ways, Fig. 8.2. If oxygen from the atmosphere is readily available 'aerobic' processes predominate but in a compact mass of organic material free oxygen is rapidly used up and, if further oxygen cannot penetrate, anaerobic processes occur in which the organisms obtain oxygen chemically combined in the decaying material. The products of anaerobic decomposition are therefore reduction compounds like methane, CH_4, together with ammonia or ammonia-like compounds, hydrogen sulphide, etc. These are evolved as gases, some of unpleasant odour.

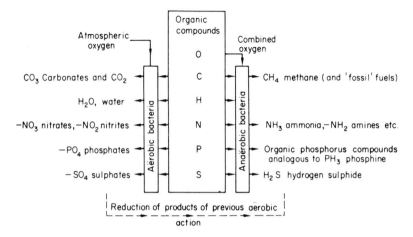

Fig. 8.2 Aerobic and anaerobic biodegradation

Water normally contains atmospheric oxygen in solution and aerobic processes are therefore possible in an aquatic environment. If untreated sewage
is discharged to a river, the dissolved oxygen is depleted to some extent by the
aerobic action, but provided re-aeration, encouraged by the turbulence of
the flow, is adequate the dissolved oxygen may not be completely removed
and after a time, when oxidation is complete, the normal concentration will
be recovered. This is shown in Fig. 8.3, curves A and B. If the polluted load
is large in relation to the size of the river as in curve C, dissolved oxygen will
be completely removed at some stage and anaerobic action will take over.
The obnoxious conditions will be aggravated by the death of all living things
needing oxygen including, of course, fish. Suspended organic matter in the
sewage will settle out so that, even if aerobic conditions are maintained in

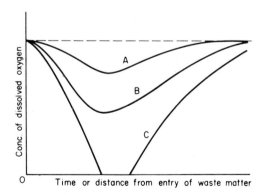

Fig. 8.3 Oxygen (sag) curves

the main part of the flow, anaerobic conditions may occur at the bed.

Conditions in a river receiving sewage may be unsatisfactory even if aerobic action is fully maintained as the lower level of available oxygen will affect the natural balance of river life.

8.3 Treatment of sewage before discharge to a river

The first method adopted for treatment was to allow the sewage to percolate through the natural soil to which it was applied by various methods of irrigation. The small numbers of aerobic bacteria and associated organisms present in the soil and in the sewage were able to multiply because of the availability of nutrients from the sewage and of oxygen, but very large areas were needed to avoid the risk of anaerobic conditions due to accumulations of solid material.

An advance was made by using beds of stone rather than natural soil. These could be contained within tanks so that treated sewage could be deliberately collected. As the beds could be several feet deep, and had ample void spaces for the penetration of air, the area needed was reduced. The beds were first operated on a batch system, being fully charged with sewage for a time and later emptied and left for aeration to occur. These were known as contact beds but were rapidly superseded by percolating biological filters in which sewage sprinkled over the surface of a bed of stones trickled downwards continuously, leaving part of the void space for air.

Percolating filters were installed extensively in the early years of the present century, though contact beds were still being installed in some places, and land treatment remained fairly common for smaller towns.

The organisms which perform the oxidation process in filters occupy a slimy layer which builds up on the stones during the first few weeks of operation. Many different species of bacteria are present, together with protozoa, fungi, and higher forms of life such as worms and insects. These compete for food and some forms prey upon others so that a balance is set up. The grazing forms are useful in that they limit the growth of the slime and so prevent the bed from becoming choked. The object in operating a filter is to maintain an appropriate balance. Applying sewage at too high a rate causes too great a development of slime and 'ponding' occurs. It is the supply of nutrients rather than the actual flow of sewage which is important. A good deal of the organic matter in sewage is present in suspended form, so it has always been customary to remove the readily settleable material in sedimentation tanks before applying the sewage to filters.

Since the purifying organisms in a filter are building up cell material from the nutrients in the sewage, there must be a net flow of solids from the filter if it is to operate continuously. Some solid material is removed by insect life, but the surplus becomes detached, possibly with the aid of the higher forms of life, and passes out with the purified liquor from which it must be removed by settlement. Humus tanks are provided for this purpose.

During the period when filters were being developed and installed, a number of workers attempted to evolve a method of sewage treatment based on the aeration of sewage in tanks. Success was achieved by Ardern and Lockett in 1913.

Merely blowing air through sewage results in very little oxidation. Ardern and Lockett[1] found that, by decanting off liquid and replacing it with fresh sewage at intervals, the sludge retained in the vessel gradually developed into a culture capable of oxidising organic material in the sewage in a period of a few hours. The sludge so formed was named *activated* sludge and apart from its oxidising capabilities it has the important attribute of being flocculent and can be readily removed from the treated liquor by settlement.

In an activated sludge plant, sewage (usually after sedimentation tank treatment) is mixed with activated sludge at the inlet end of the aeration tank. The mixed liquor is agitated and aerated either by air bubbles or by surface paddles. Fairly violent agitation is needed to promote mixing of air and sludge with the waste and to keep sludge in suspension. Sewage and sludge flow continuously through the tank which must be large enough to accommodate a sufficient stock of micro-organisms at a concentration low enough for adequate aeration. After this the mixed liquor passes into final settling tanks where the activated sludge is settled out and the clarified liquid passes over an outlet weir for discharge to the river. The activated sludge is continuously removed to minimise loss of activity and returned to the aeration tank inlet for reuse.

The mass of activated sludge increases during the process, due to the conversion of sewage organics to cell material. The surplus is treated along with solids settled from the raw sewage before oxidation.

Though there are similarities in the biology of filters and activated sludge, there are some significant differences. In the filter, since the biological material remains stationary, a degree of specialisation is possible at different levels of the bed and the population of organisms changes with depth. In the lower regions, organisms which thrive on partly oxidised waste are common and can take the oxidation to a further stage. There is less opportunity for this in activated sludge since the organisms move with the water. The environment of an aeration tank is thoroughly aquatic so many of the higher forms of life, such as insects, associated with filters do not occur.

In practical terms the main contrast between filters and activated sludge plants is that the latter are cheaper in capital cost and occupy much less space, but are more expensive to run because of power needs. More detailed comparison is made in Chapter 11 where a number of variations of both processes are described.

An alternative to filters or activated sludge is the algal pond. Algae are able to use solar energy to perform photosynthesis and thereby to provide oxygen. Bacteria, associated with other organisms, still perform most of the purification but algae provide an important source of oxygen. Since the ponds have to be of large extent, however, aeration of the water surface by atmospheric oxygen plays a part also. Anaerobic processes occur in the lower levels of the pond and in some places throughout at some seasons. Algal ponds are perhaps most common in hot sunny climates, but they are in use as far north as Alaska. Design criteria differ with climate.

In outlining the biological processes emphasis has been laid on the stabilisation of organic matter so that its demand for oxygen will be satisfied before discharge. Sewage contains material other than waste of organic origin. Many substances which are organic chemically but derived synthetically are degraded to some extent in the normal process. Some other constituents are readily removed at some stage of treatment. These include gross solids which can be screened out, grit which settles out, grease and oil which can be skimmed off, and dissolved gases which are simply released to the atmosphere.

The primary object in introducing the water-carriage system of waste removal was the prevention of water-borne diseases. Typhoid, paratyphoid, cholera and some forms of dysentery are caused by bacteria and are contracted by ingesting water or food which has been in contact with the faecal matter of sufferers from the disease or, in the case of typhoid, of carriers. Some virus diseases may be transmitted in a similar manner.

The prevention of water-borne epidemics of these diseases is secured by the efficient removal of faecal matter, which a proper sewerage system ensures, together with the supply of clean or disinfected water through mains continuously under pressure and as near leak-free as possible. No deliberate steps are needed in sewage treatment to kill the organisms causing these diseases since a sewage-treatment works is an unattractive environment for pathogenic bacteria, whose natural habitat is at body temperature. Occasionally, where it is imperative to kill all pathogenic bacteria before discharging an effluent to a river, chlorination is adopted. The use of indiscriminate bactericides should always be carefully considered, however, as they will kill the organisms responsible for natural purification processes as well as the pathogens.

For a sewage works treating domestic sewage or biodegradable trade wastes the main criterion of performance is the reduction in the oxygen demand of the flow to a level which can be tolerated in the river. However, the increasing use of sewerage systems for the reception of trade wastes has directed attention towards the need for additional criteria and towards the means of achieving them. There are sewage works producing effluents which are satisfactorily low in oxygen demand but which contain colouring matter, turbidity, foam, material toxic to river life and substances which render river water unsuitable as a source of water supply. A possible remedy is to apply further processes of treatment to the effluent ('tertiary treatment', see Chapter 13).

Another possibility is to introduce physical or chemical methods of treatment, for example the use of coagulants in primary sedimentation, and the use of activated carbon to absorb organic matter from solution. The use of semi-permeable membranes offers another alternative. So far none of these processes appears to give an economic alternative to the conventional processes for municipal sewage treatment but one of them may be the solution to the treatment of a particular trade waste.[2,3]

8.4 The complete sewage works

Biological oxidation is the main method of stabilising the organic constituents of liquid waste and can be used for both dissolved and suspended impurities. All the processes are relatively costly in one way or another and in most cases it is economic to reduce the load on the biological oxidation plant by removing settleable solids first. Of course this leaves a sludge to be dealt with but, though sludge treatment and disposal is often regarded as an expensive and intractable problem, it is often cheaper to use one of the common methods of separate sludge disposal than to dispose of sludge by biological oxidation along with the dissolved and colloidal matter. Biological oxidation of sludge requires the process to be taken to the endogenous stage and requires very long aeration periods.

The simplest way of dealing with sludge is to use it on agricultural land as a manure. Occasionally it is used in its raw state but more often it is partly dried first, by drainage and evaporation on open beds. In either case, distributing sludge on agricultural land enables biological degradation to occur over a wide area with little offence, and returns nutrients to the soil at the same time. In dry periods, the irrigation value of wet sludge is not unimportant.

The amount of sludge to be dealt with on drying beds or otherwise can be reduced by digestion, in which anaerobic processes of decomposition are encouraged by maintaining the sludge at about 30°C in closed tanks. The digestion process has the additional benefit that some of the organic matter is converted into methane which can be used as fuel in engines to provide power for the biological oxidation process.

Even without conversion into methane, organic solids in sludge are combustible. If sufficient water can be removed first, and the temperature is high enough, combustion of finely divided sludge can be maintained without any other fuel. This method of sludge disposal which leaves only a small quantity of inorganic ash is fairly new and not yet in extensive use.

It should be noted that oxidation is a common factor in a number of alternative means of treating organic waste material in sewage. The dissolved and colloidal fraction is oxidised biologically, the suspended solids fraction can be oxidised biologically but part or all of it is sometimes oxidised by combustion, often after using an anaerobic process to render part of it more combustible. Some of the dissolved and colloidal matter is ultimately dealt with in this way in that it becomes humus or surplus activated sludge. With the processes available at present, it seems to be economic to restrict biological oxidation to the dissolved and colloidal impurities but there are few reasons other than those of economy against striking the balance at some other point.

Whether or not sewage is to be treated in sedimentation tanks before oxidation, gross solids and inorganic grit have to be dealt with first. The former are arrested by some form of screen and are shredded and returned to the flow or disposed of by burying or incineration.

Grit is easily removed by settlement in ˌelatively small tanks and if it is washed free of organic material it is innocuous enough to be used as filling in footpath construction. If grit were not removed before organic material were settled, the sludge would be more difficult to handle. In the few cases where suspended organic material is not settled out before oxidation it is still better to eliminate grit before treatment.

Sewage treatment works have to be capable of dealing with wide variations in flow. In dry weather the variation is typically from one-third of the mean during the night to twice the mean during the middle part of the day. In combined sewers the wet weather flow would reach 50 or even 100 times the mean if storm overflows were not provided to spill a large proportion of the maximum flow without treatment.

In the U.K. it has always been considered to be uneconomic to provide full treatment capacity for flows in excess of about three times the dry weather mean (see page 311 for a more precise description). Sedimentation tanks and bio-oxidation stages of treatment are normally designed to take flow up to this maximum.

Storm overflows on combined sewerage systems are set to pass to the treatment works flows up to a maximum which is about twice the full treatment capacity (see page 84). The whole of this flow is given screening and grit removal treatment but flow during storms in excess of the main line capacity is diverted to storm tanks. These tanks remain empty during dry weather and fill completely only in the more severe storms. They then overflow to the river but since they are designed to operate as sedimentation tanks, the discharge receives partial treatment. After each storm the tanks are emptied by pumping their contents into the main line where full treatment is given.

A conventional sewage-treatment works consists of the following stages.

1 Preliminary Treatment: Screens. Grit (or detritus) settling tanks. Flow measurement. Storm overflow weir.

2 Primary Treatment: Primary sedimentation tanks. Storm tanks or balancing tanks in parallel with primary sedimentation tanks. Storm- or balancing-tank emptying pumps.

3 Sludge Treatment: Physical or biological de-watering processes. Transport of sludge from site for disposal, or incineration.

4 Secondary Treatment: Biological oxidation by filtration or activated sludge, occasionally in combination, This stage includes separate tanks for humus or activated-sludge settlement which are an essential part of a biological oxidation process, Fig. 8.4.

To this might be added tertiary treatment which would include micro-strainers or sand filters to improve the effluent by removing further suspended organic material. Slow sand filters perform some oxidation as

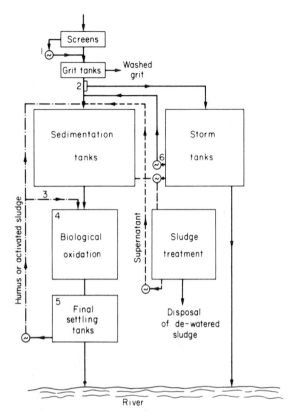

1. Disintegrator shreds screenings
3. Returned activated-sludge pipe
   (not needed in filter plant)
5. 'Humus' tanks on filter plant

2. Storm weir (flow>3 x DWF
   overflowed to storm tanks)
4. Aeration tanks (activated sludge)
   or percolating filters
6. Empties storm tanks after
   each storm

Fig. 8.4 Flow sheet for conventional sewage treatment

well. Chlorination would also be included under this heading.

The above classification arises from the idea that the stage to which treatment is carried depends on the water receiving the effluent. Thus a town discharging sewage to the open sea might provide preliminary treatment only; discharge to a large river or estuary might not require treatment beyond the primary stage. There are few places in Britain where the secondary stage is not justifiable and in some tertiary treatment is needed as well. Relatively arid climates where water is precious and has to be reused also demand tertiary treatment of sewage.

This classification is too rigid to be perfectly general. It has been seen already that it is possible to treat unsettled sewage by biological oxidation and thereby to eliminate primary settlement. Pre-aeration of sewage before settlement is sometimes of value and a further step is to use activated sludge to absorb suspended solids at an early stage, to settle out the sludge and aerate it in a separate tank while treating the supernatant liquid in filters. It is difficult to fit such expedients into the conventional primary–secondary–tertiary concept though they are quite practicable.

8.5 Laboratory tests

Brief details of the most common tests[4] and of their interpretation are given below. These represent the bare minimum of knowledge which a design engineer should possess. Engineers intending to specialise in sewage-works design benefit from a period of a few months in a sewage-works laboratory, particularly if they have studied chemistry to 'A' level standard.

Sampling

Because the flow and strength of sewage vary throughout the day and with rainfall, an individual 'spot' sample is of little value. For this reason 'composite' samples are often preferred. These are obtained by mixing samples taken at regular intervals, say hourly. The samples may be mixed in equal proportions or, preferably, in proportions corresponding to the flow rates at the times of sampling. Composite samples give an average impression of the sewage or effluent. Though individual spot samples are of little value, series of spot samples taken at regular intervals and tested separately are of more value than composite samples. The time interval between samples should be carefully considered and preliminary investigation by taking samples at various intervals of time is worthwhile. This will usually indicate that, for raw sewage, samples need to be taken at intervals appreciably shorter than one hour during the day if a faithful picture of the variations is to be obtained. At intermediate and final stages of treatment, less frequent sampling is needed as the processes smooth out the fluctuations.

Samples may be taken by dipping a can into the flow and filling sample bottles supplied by the analyst. The bottles need referencing and records are kept of the date, time and place of sampling and of the rate of flow. Automatic samplers are used for much routine work. A small pump continuously draws the liquid to be sampled. The flow is diverted to bottles, often on a rotating table, at fixed intervals of time. When the flow is not discharging to a bottle it runs back to the source.

Some of the properties customarily tested can now be monitored continuously by passing a small flow through specially designed instruments.[5]

Analytical results used in connection with design are usually taken from samples obtained during dry weather. The justification for this is that the complicating factor of dilution by rainwater is avoided but critical design cases may exist between the extremes of high flow of low strength and low flow of high strength.

Tests on solid contents

Solids are present both in solution and suspension.

The concentration of total solids is found by evaporating a measured volume of sample to dryness, and weighing the residue. Evaporation is done at 100°C to 105°C on a water bath to avoid chemical reactions which might occur at higher temperatures.

Several methods are available for suspended solids determination. A measured volume may be filtered through a specially prepared asbestos mat in a Gooch crucible. The mat is dried (at 100°C–105°C) and weighed. It is now more common to separate the solids by centrifuging in two stages and then to dry them for weighing. A third method is to filter the sample through a glass-fibre disc in a Buchner funnel. The Gooch crucible method gives higher results than the centrifuge method since some colloidal material is retained on the mat. The glass-fibre filtration method gives results comparable with those of the Gooch crucible test and is frequently used.

The liquid separated in the above test can be used to determine dissolved solids by evaporation. The suspended solid matter from the centrifuge test can be ignited to determine the proportions of volatile and mineral matter. Visual observation during ignition gives qualitative information and the residue can be tested qualitatively for inorganic constituents.

The concentration of suspended solids gives an indication of the amount of matter which will be removed in sedimentation tanks or which, in a final effluent, might settle out in the river giving obnoxious conditions. It is no more than a rough guess because the analytical methods are more efficient than settlement processes. A settleable solids test gives a better indication but is less precise. A sample is placed in a vessel, often an inverted (Imhoff) cone with graduations and the volume of sludge is measured after a fixed time of a few hours. For research or development purposes, water-jacketed or heat-insulated settling columns are sometimes used with draw-off points at various levels. The column is filled with sewage and the samples taken at intervals from the draw-off points. Solids determinations are carried out on the samples and the results give an analysis in terms of settling velocity of the suspended solids in the original.

As the liquid is quiescent in these tests, the conditions in a settling tank are not completely reproduced.

The volatile solids concentration gives a rough indication of the amount of organic matter present and therefore of the strength of the sewage.

An indirect assessment of suspended solids content can be obtained by measuring turbidity (cloudiness) or the complementary property of transparency. Turbidity can be measured by visual comparison or by photo-electric instruments in relation to an arbitrary standard, e.g., a suspension of fuller's earth. The presence of colour is a complicating factor. Transparency is measured by slowly filling a glass tube until a mark at the bottom becomes invisible. Klein[6] used a tube 2 ft long by 1 in. diameter with a mark in the form of a cross with mm lines and measures transparency in mm of liquid in the tube. The test is performed on both an original sample and on the supernatant part of a settled sample and the difference is said to give a good indication of suspended solids content.

Oxidation tests — Biochemical oxygen demand

The sample is diluted with aerated water and divided into two portions. The dissolved oxygen is determined in one sample immediately and in the other after it has been incubated for 5 days at a temperature of 20°C. The difference gives the amount of oxygen taken up by the sample.

Oxidation is through the agency of aerobic bacteria present in the sample and the BOD test is often a better guide to necessary oxidation capacity than chemical tests in which oxidation is by direct action. The presence of toxic material results in low BOD values, however, and many other factors complicate interpretation. Thus the nature of the diluent water has an effect and in some cases the degree of dilution. Also the test is sensitive to the temperature of incubation. Misleadingly high figures can be obtained if the sample has reached an advanced stage of oxidation in which compounds simpler than the original carbonaceous matter are being oxidised.

The biological oxidation of nitrogenous compounds occurs in the later stages of the process and may or may not have started during the standard 5-day period. It is likely to be more advanced in samples from later stages of treatment. In order to remove this complication in interpretation of BOD values, nitrogenous oxidation can be suppressed by adding a small quantity of allylthiourea (ATU) to the sample.

Though one of the most useful tests in sewage laboratory practice, BOD can be the most difficult to interpret.

Oxidation tests — Permanganate value (PV)

The sample is mixed with an acidified solution of potassium permanganate in a stoppered bottle and left at a temperature of 27°C. Oxidisable matter reacts with the permanganate. The amount of oxygen absorbed is deduced at the end of the test from the amount of permanganate left, determined by using it to liberate iodine from sodium iodide, and titrating with sodium thiosulphate. Two tests are normally performed, one after 3 minutes and another after 4 hours.

Provided that no oxidising agents are already present in the sewage, the tests indicate oxygen demand and give a rough idea of the extent of oxidation treatment required. The tests give also an indication of the amount of oxidisable matter present though oxidation is often incomplete.

The 3-minute test indicates the amount of readily available oxidisable matter, probably inorganic and perhaps arising from trade waste. The 4-hour test indicates the less readily oxidised matter as well. The ratio of the two values is of interest when trade wastes are involved.

For sewage the ratio 4 h:3 min PV is about 3:1, for vegetable processing waste, 4:1 to 10:1 while for gas liquor the ratio is about 2:1.

If abnormally low values (they are occasionally negative) are obtained and the presence of oxidising agents is suspected, their effect can be eliminated by carrying out a preliminary test without permanganate to determine the permanganate equivalent of the agents present.

Oxidation tests — Chemical oxygen demand (COD)

In this test potassium dichromate is used as the oxidising agent. It is preferred because organic matter is more completely oxidised than in the permanganate test. Also it is quicker to perform and is now often used as a routine monitoring test after establishing the ratio of BOD and COD for the particular waste. The COD is about double the carbonaceous BOD for settled domestic sewage but higher ratios, of five or more, may be found for final effluent. The presence of oxidising or reducing agents and chlorides complicates the interpretation of COD results. Ammonia is not normally oxidised in the test so the nitrogenous part of the oxygen demand is not included in the COD.

Oxidation tests — Total oxygen demand (TOD)

This test employs catalytic oxidation at $900\,^\circ C$ in a combustion tube, using a very small sample and giving a result in 5 minutes. Both carbon and nitrogen compounds are oxidised almost completely. However, in interpreting the results regard must be had to the effects of oxygen already present in the sample, for example in nitrite or nitrate form. Unlike the COD test, TOD can be used for samples of high salinity.

Total organic carbon (TOC)

The carbon in the sample is oxidised to carbon dioxide in a catalytic combustion chamber or by ultra-violet light and the carbon dioxide determined by an infra-red or flame ionisation technique. The whole procedure is automatic and a number of patterns of instruments are available, even offering the facility to determine organic carbon in dissolved, suspended and volatile form. The TOC concentration is about one third of the BOD.

Nitrogen tests

Waste matter in sewage consists in part of organic nitrogen compounds. These are broken down during treatment to ammonia and related compounds which are oxidised to nitrous ($-NO_2$, nitrite) and ultimately nitric (– NO_3, nitrate) nitrogen. Thus the total concentration of nitrogen present is a good indication of sewage strength and the relative proportions of the different forms of nitrogen indicate the stage of oxidation reached.

'Ammoniacal nitrogen' is a measure of that present as free and saline ammonia. 'Albuminoid nitrogen' is that part of the organic nitrogen which is converted to ammonia by alkaline potassium permanganate. Nitrogen in nitrous and nitric forms is determined in addition though little is to be expected in a raw sewage. A fully oxidised effluent should contain little ammoniacal, albuminoid and nitrous nitrogen but will contain significant nitric nitrogen oxidised during treatment from the proteins and amino acids in the waste.

Stability tests

A stable effluent is one which is likely to remain aerobic. For this to be so, oxygen must be available in solution or in the form of nitrate, sulphate, etc. The chemical situation in the effluent will not be static. Both oxidants and reductants will be present in a dynamic equilibrium. This equilibrium may be measured by the value of the oxidation-reduction or 'redox' potential based on an electronic interpretation of the reactions. If the value is positive, oxidation dominates, if it is negative, reduction. There is therefore a close relationship between aerobic and anaerobic biochemical processes and redox potential.

The dye methylene blue is converted to a colourless compound by reducing agents and the loss of colour occurs at a redox potential of approximately zero. It is therefore a useful indication of stability in a sewage effluent. In testing, a small quantity of methylene blue is added to the sample which is then kept out of contact with air and incubated at 20°C for 5 days. The concentration of methylene blue must not exceed 1.33 mg/l as higher concentrations are bactericidal. The time required for decoloration to occur is a measure of the instability of the sample. If decoloration has not occurred after 5 days, the sample is regarded as being stable.

There are several alternative tests, for example, 3-minute permanganate tests may be carried out before and after 5 days incubation at 80°F. An increase in permanganate value indicates instability.

Other tests

Synthetic detergents in sewage cause foaming troubles in sewage works as well as in rivers and they reduce the rate of transfer of oxygen into solution. Problems have been alleviated by the use of 'soft', biodegradable detergents (like alkyl sulphates) in place of 'hard' less degradable detergents (like alkyl aryl sulphonates) formerly popular. Hard detergents are still used industrially to some extent. Foaming in aeration tanks can be controlled by continuous dosing with small quantities of a suppressing agent.

The presence of trade wastes may necessitate a great variety of tests to check whether unusual problems will be created. These tests may seek toxic material, both organic and inorganic, metallic and non-metallic; fats, grease, oils, and radioactive isotopes. The primary source of information about trade waste will usually come from analysis of the wastes themselves before discharge to the sewers but it is usually advisable to investigate the sewage as well to obtain a check on the overall position, especially if unauthorised discharges are suspected or interaction is possible.

General interpretation of test results

From the brief remarks about interpretation of results of individual tests it will be realised that each type of test cannot be considered in isolation. Thus comparisons of BOD and PV results may reveal inconsistencies, and volatile solids, for example, or results from tests for toxic substances may help to explain these. It is necessary to examine the results from all the different tests, and extensive experience is required in order to report usefully on the extent and kind of treatment required.

8.6 Typical analytical results

Mean concentrations of BOD, permanganate value and suspended solids in raw sewage and after primary and secondary treatment usually lie between the limits shown in Fig. 8.5.

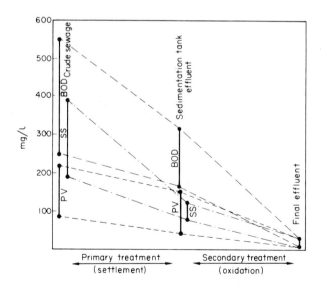

Fig. 8.5 Changes in biochemical oxygen demand, permanganate value and suspended solids concentrations during treatment, showing normal ranges

The values obtained in a particular case depend on the consumption of water, on the presence of trade waste and on the diet of the population. For domestic sewage in Britain, BOD amounts to about 0.055 kg per head of population per day. With water consumption of 200 l/head-day this gives a BOD value of $0.055 \times 10^6/200 = 275$ mg/l.

Primary sedimentation tanks, by removing 65% to 75% of the suspended solids, reduce the BOD of the liquor by 30% to 35%. The removal of suspended solids and BOD in primary sedimentation is lower for weak than for strong sewage.

The oxidation stage, whether employing filters or activated sludge, reduces both BOD and suspended solids to very low levels. At the same time the form in which nitrogen is present changes, ammoniacal and albuminoid nitrogen decreasing while the oxidised forms, nitrous and nitric nitrogen, increase. With single-pass filters, a high degree of conversion to the nitric form can be expected but, with filters operated at the higher rates possibly with recirculation of effluent or alternating double filtration, the effluent is usually less well nitrified. Though activated sludge plants produce a good deal of nitrification in favourable circumstances, they do not usually nitrify so well as single-pass filters.

Table 8.1 shows results from the Rye Meads works in which a form of tertiary treatment is included.

8.7 Effluent quality

The standard of quality of effluent to be aimed at in treatment works design depends upon the river or other body of water which is to receive the discharge.

In the case of a river which is still almost in its natural state, supporting fish life and possibly used as a source of water supply, the standards will need to be fixed to preserve its condition. For rivers which are already polluted there will be a long-term aim to restore them if at all feasible so that new works will often have to be designed to produce a high-quality effluent ultimately if not immediately.

The water authorities have the difficult task of setting up standards for river-water quality and consequently for discharges into the river, taking into account both the use of the river as a source of water supply for various purposes and its use for conveying residual waste products to the sea. Rate of river flow is an important factor since it affects the dilution which discharges receive. It will vary seasonally and also randomly and will be affected by abstractions, discharges and the effects of reservoirs.

Organic waste matter is not simply taken away by the river. If the river is well aerated the waste matter will be 'purified' to a further degree by aerobic processes. If the river is not well aerated, there will be a risk of anaerobic conditions arising and the rate at which waste matter can be safely discharged will be very much smaller.

Table 8.1 Analytical results in parts per million (mg/l) for Rye Meads works showing progress of treatment. The results are average for the period Nov. 1958 to April 1959. (From Balfour, D.R., Manning, H.D., Cripps, T. and Drew, E.A., The design, construction and operation of the Middle Lee Regional Drainage Scheme, *Proc. Inst. Civ. Engrs*, **17**, 283–314, Nov. 1960)

	Permanganate value		5-day BOD	Nitrogen				Suspended solids	Syndets	pH	Alkalinity	Chlorides
	3 min	4 hour		Ammoniacal	Albuminoid	Nitrous	Nitric					
Crude sewage	25.2	77.5	322	31.6	13.6	0.32	1.2	380	17.1	7.26	425	96
Sedimentation tank effluent	15.4	44.5	195	31.2	7.2	0.58		121	10.0	7.34	414	94
Final tank effluent	3.7	10.9	18.6	10.6	1.5	0.66	15.2	12.1	5.3	7.33	278	94
Rapid gravity sand filter effluent	3.2	9.4	8.7	11.1	1.2	0.37	15.6	4.0	4.3	7.39	286	94

The two most important, easily measurable, factors relating to effluent quality are the concentrations of suspended solids and of 5-day BOD. The suspended solids left in a sewage-works effluent are largely organic and if there is an excessive amount, deposition and subsequent anaerobic conditions may occur in the river. The BOD value is a measure of the demand placed on the river for further aeration. A part of this will be the oxygen demand of the suspended matter, so the two factors are not independent.

Nor are these the only factors which might have to be taken into account. Toxic materials such as metallic ions and synthetic organic compounds will have to be limited if there is any risk of their being present. They may have to be limited in raw sewage, perhaps by prior removal, to prevent interference with treatment processes, but treatment processes are sometimes more tolerant than river life, as acclimatisation is sometimes possible and control can be exercised.

Thus toxic material in concentrations high enough to cause concern may exist in the effluent from a biological process. Even substances which are not toxic in any likely concentrations (for example synthetic detergents) may impair the quality of the river water for some purposes.

The form in which nitrogen exists in the effluent may be important. A good deal of the nitrogen in raw sewage appears in the final effluent from a conventional treatment works but it will have been oxidised to a varying degree from ammonia and related compounds to the nitrite or nitrate form. Nitrogen which is still in the form of ammonia will impose an oxygen demand in addition to that measured as BOD, will require more chlorine for disinfection when water is abstracted for public supply and can be toxic to fish when the pH is also high. Different biological oxidation processes producing comparable reductions in BOD do not necessarily produce the same degree of nitrification (i.e. conversion of nitrogen to the nitrate form). As mentioned above, conventional percolating filters produce well-nitrified effluent while activated-sludge processes, though achieving comparable BOD reduction, do not produce such well-nitrified effluents unless they are generously designed and operated.

In Britain the standard of quality of sewage effluent discharged to a river has commonly been taken as 5-day BOD not to exceed 20 mg/l and suspended solids not to exceed 30 mg/l. This is known as the 20/30 standard and derives from a recommendation in the Eighth Report of the Royal Commission on Sewage Disposal published in 1912. It was associated with the assumption that the river would afford a clean water dilution of at least eight times.

The quality of sewage works effluent varies due to changes in load on daily, weekly and seasonal cycles, due to rainfall and due to changes in operation for maintenance and other purposes. Breakdowns, industrial disputes and other occasional happenings may also affect effluent quality. Typically[7] the variation is such that about 5% of samples have more than twice the mean concentration of BOD or suspended solids, and about 5% of samples have concentrations less than half of the mean values. Mean concentrations of BOD and suspended solids may therefore need to be as low as 10 and 15 mg/l respectively if the 20/30 standard is interpreted as a 95% confidence limit. Regarding 20/30 as an absolute standard implies still lower mean concentrations. However, it must be recognised that the possibility of exceeding a set limit can never be entirely eliminated.

The 20/30 standard can be fairly readily achieved with a conventional treatment works employing primary settlement, an oxidation stage and final clarification. For this reason it is likely to continue in use as a norm.

Where higher standards (that is lower concentration) than 20/30 have to be achieved, further final clarification is often all that is needed. Micro-strainers or sand filters can be used to remove suspended matter from the humus or activated-sludge settling-tank effluent. The reduction of suspended solids concentration by these processes automatically reduces the BOD of the effluent by about 1 mg/l for every 3 or 4 mg/l reduction in suspended solids. Thus standards higher than 20/30 could be, for example, 16/22 of 12/14, though as no great precision is warranted standards of 15/20 or 15/15 would be considered more practical. Either of these might imply that a tertiary treatment stage of suspended solids removal was needed though there are works producing effluents of this standard without tertiary treatment.

More extensive biological treatment would be needed only if there were a further requirement to produce a well-nitrified effluent. In that case, possible expedients would be to use single-pass percolating filters of fairly large capacity, a large activated-sludge plant or higher-rate filters following an activated-sludge plant. With some sewages, sufficient nitrification could be achieved with an activated-sludge plant, only by subsequent biological filtration.

Though in Britain the main concern is to preserve or restore life in natural waters, it should not be forgotten that the discharge of sewage effluent provides nutrients which in some circumstances can cause water life to develop to an undesirable extent. Algae and rooted plants may become a problem. Processes have been developed for removing the main nutrients, nitrogen and phosphorus compounds. Nitrogen can be removed by an adaptation of the actived-sludge process and phosphorus by a physical ion-exchange process. Significant supplies of nutrient reach rivers from agricultural operations.

Where quality standards lower than 20/30 are acceptable, the capacity of the biological oxidation stage can be reduced. In principle, though not always in practice, a percolating filter or activated-sludge process can be proportioned to give an effluent of any desired BOD between that of the raw water at one extreme and a BOD of the order of 10 to 20 mg/l at the other. If the biological or secondary stage is omitted altogether and treatment consists of settlement alone, about 75% of the suspended solids can be removed and there will be a consequent removal of perhaps 35% of BOD. A sewage of 300/400 might be treated to give an effluent of 200/100. In Britain an effluent of such high strength would be acceptable only for discharge to tidal waters except for occasional discharge of storm-tank effluent. If partial biological oxidation were adopted, the final settling tanks would be of much the same size as for full-scale oxidation, since the delivery of suspended solids from filters is intermittent and varying, while the final tanks of an activated-sludge plant have the important function of recovering activated sludge for reuse. Consequently, in such a case, a suspended solids standard would be almost irrelevant and BOD would be the main consideration.

8.8 Performance formulae for sewage treatment design

Until quite recently capacities of treatment stages have been based mainly on rules-of-thumb many of them deriving from Royal Commission recommendations and being slowly modified in the light of experience and in response to developing ideas and equipment.

Thus, the detention period for primary sedimentation was gradually reduced to about 6 hours DWF as it was found, for example at works which were becoming overloaded, that reduction in detention time led to only a small reduction in performance, especially after the introduction of mechanical scrapers enabled tanks to be de-sludged more frequently.

It was known that tanks of this size removed from the flow about two-thirds of the suspended solids and thereby about one third of the BOD. This information was sufficient to enable estimates to be made of the strength of sewage passing to the next stage of treatment, the design of which would also be based on rules derived from experience.

Further reduction in detention time of primary sedimentation would reduce the capital cost of the tanks and the amount of primary sludge produced but would leave more BOD to be removed in bio-oxidation and some of this BOD would be converted into secondary sludge, which is more difficult to de-water than primary sludge. This is but one example of the inter-relationships of treatment stages. The optimisation of the design decisions for a whole works to achieve least cost became feasible, in principle, once large computers were available. Various studies have been made.[8,9,10]

Essential to a cost optimising model are mathematical expressions enabling the output of each process to be calculated from the input according to the size of the unit. One of the benefits of the optimisation studies has been the development of performance functions and these can now be used to carry out routine design calculations more rationally.

However, no great precision can be expected in this field. The properties of sewage vary both from place to place and from time to time in the same place. A further difficulty in fitting mathematical expressions or validating theory arises from the concentration dependence of many of the processes resulting in performance as shown in Fig. 8.6, where the final concentration is insensitive to detention time.

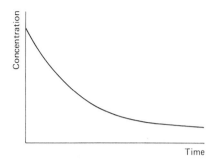

Fig. 8.6 Concentration dependent process

8.9 Discharge of sewage to tidal waters

The discharge of sewage to estuaries or to the sea at a short distance from the shore is extremely common. Frequently, the only treatment given is screening, the macerated screenings being returned to the flow. Float tests are carried out over a range of tidal conditions to aid in siting the outfall and to discover at what states of the tide discharge should be avoided. Storage tanks are needed for periods when discharge has to be prevented.

Many of these schemes have resulted in the pollution of beaches. The health hazard appears to be negligible but aesthetic objections are quite sufficient to condemn the discharge of untreated sewage to the sea except in deep water at a long distance from the shore.

The development of marine pipe-laying techniques in the oil industry, and the necessity to find means of conveying radioactive wastes to deep water have resulted in long sewage outfalls becoming feasible. The pipelines usually require multiple outlets, and a good deal of work has been done on the design of these to secure diffusion and mixing of the effluent.[11,12,13]

The acceptable alternatives in discharging to tidal waters are treatment before discharge, or a deep-water outfall. Where treatment is adopted, effluent standards need not be so stringent as for discharge into fresh water. In some cases primary treatment might be sufficient. At a large plant, it might be found economic to use a deep-water outfall for sludge disposal.

References

1 Ardern, E. and Lockett, W.T., Experiments on the oxidation of sewage without the aid of filters, Pt. I, *J. Soc. Chem. Ind.*, **33**, 523–539, 1914. Reprinted in *J. Inst. Sew. Purif.*, Pt. 3, 175–188, 1954.
2 Cooper, P., Physical and chemical methods of sewage treatment — review of present state of technology, *Wat. Pollut. Control*, **74**(3), 303–311, 1975.
3 Banks, N., Research and Development in the U.K., *Wat. Pollut. Control*, **74**(3), 312–327, 1975.

4 Department of the Environment, *Analysis of Raw and Potable Waters*, H.M.S.O., 1972.
5 Briggs, R., Schofield, J.W. and Gorton, P.A., Instrumental methods of monitoring organic pollution, *Wat. Pollut. Control*, **75**, 47–57, 1976.
6 Klein, L., *River Pollution,* Butterworth, London, 1966.
7 Porter, K.S. and Boon, A.G., Cost of treatment of waste water with particular reference to the river system of the Trent area, *Proc. Trent Research Programme Symposium*, Inst. Wat. Pollut. Control, 1971.
8 Water Research Centre, Sewage treatment optimising model.
9 C.I.R.I.A., Cost-effective sewage treatment — the creation of an optimising model, *Report No. 46*, May, 1973.
10 Tarrer, A.T., Grady, C.P.L., Lim, H.C. and Koppel, L.B., Optimal activated sludge design under uncertainty, *Proc. Am. Soc. Civ. Engrs*, **102**, EE3, 657–673, June 1976.
11 Inst. Public Health Engrs, Sea outfalls symposium, in *Public Health Engr*, **12**, 89–114, April 1984. Papers by: M.W. Knill, K.J. Flemons and C. Mason; J.P.N. Mackenzie; J.K. Reynolds; and D. Munro.
12 Grace, R.A., Sea outfalls — a review of failure, damage and impairment mechanisms, *Proc. Inst. Civ. Engrs*, Pt. 1, **77**, 137–152, Feb. 1985.
13 Mason, C., Flemons, K.J. and Taylor, A.G., Planning, design and construction of the Great Grimsby sewage outfall, *Proc. Inst. Civ. Engrs*, Pt. 1, **78**, 1045–1064, Oct. 1985.

9

Preliminary treatment

9.1 Screens

The spacing of the bars in main screens is a compromise. From the point of view of subsequent treatment, close spacing would be advantageous but the screen would then be easily blocked and difficult to clean. Commonly the space between the bars is 20 to 25 mm. Occasionally 50 mm spaces have been used but this allows too much material to pass through.

It is common to use 10-mm thick bars where the spaces are 20 mm and 12.5-mm thick bars for 25 mm spaces so the 'blockage ratio' is usually 1:3. The bars taper in cross-section (Fig. 9.1) like fire-bars to reduce the risk of wedging.

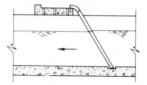

Fig. 9.1 Section of screen bars   **Fig. 9.2** Hand-raked screen

At small plants where the screens are manually cleaned, the bars slope in elevation as shown in Fig. 9.2. The screenings are raked up on to a platform spanning the channel and supporting the top of the screen. If the slab is sufficiently far above the adjacent ground, screenings may be pushed into a barrow. Otherwise, the platform needs a curb against which the screenings can be shovelled.

At most works, mechanical cleaning gear is used to reduce the labour needed. In early machines, rakes were carried on endless chains passing round sprocket wheels. The rakes travelled up the screen and then followed a path which took them above and behind the screen, where a swinging plate or revolving brush swept the screenings into a skip or, if they were to be disintegrated, into a trough or on to a conveyor belt.

Though machines of this type have often given many years of trouble-free service, machines requiring fewer submerged moving parts are now preferred.

These are of two main types. What is sometimes called the 'grab' type is used with vertical or fairly steeply sloping straight bar screens. The rake is carried on the lower member of a rectangular framework. The framework is moved up and down in a fixed guide frame by means of roller chains or wire rope. On the downward journey the rake tines are clear of the screen. When they reach the chanel floor, the framework moves forward so that the tines mesh with the screen bars on the upward journey. Screenings are cleared from the tines by a swinging plate. This description is a general one. Points of detail differ between different manufacturers.

A somewhat neater arrangement can be achieved by using a screen in which the bars form a circular arc when viewed from the side. The frame carrying the rake is then radial to the arc of the bars and is pivoted on an axis above the channel and upstream of the screen. A system of cranks and links converts the rotary motion of the driving motor to a swinging motion of the rake framework. This type of screen and cleaning gear is shown in Fig. 9.3. Another pattern, Fig. 9.4, employs hydraulic cylinders to move the raking unit.

Fig. 9.3 Dorrco Type 'T' bar screen at the Northern Sewage Works serving the Borough of Royal Tunbridge Wells (*Courtesy of the Dorr-Oliver Company Ltd*)

Where the screen-channel is deep, the grab type of mechanism is commonly used. For a channel near ground level, the mechanism associated with curved screens is less unsightly than the head frames of a grab machine.

Fig. 9.4 The Shallomatic screen (*Courtesy of Ames Crosta Babcock Ltd*)

Some raking machines are designed to reverse before the floor of the channel is reached when an accumulation of screenings is encountered so that the material is moved in successive 'bites'. Provision is also made for the tines to ride round stubborn material. Shear pins or other fail-safe devices are commonly incorporated.

It is sometimes considered desirable to provide a building for the screens and cleaning gear to protect both men and machinery from the weather. A screen house is not essential but where a good deal of attendance is needed during storms it may be well worth the extra cost.

Cleaning gear can be automatically operated by means of float-wells upstream and downstream of the screen. The floats are linked together so that they register the difference between upstream and downstream water-level. As soon as the difference reaches 150 to 200 mm due to partial blockage of the screen, the mechanism is started. Operation is continued until the difference is reduced to that for an almost clean screen. In addition to automatic control on a level difference basis, the mechanism can be arranged to operate at pre-set intervals of time or to run continuously. Provision for manual starting and stopping is of course always available.

9.2 Coarse screens

Where mechanical screen-cleaning gear is adopted it may be advisable to install coarse screens upstream to arrest large objects which would interfere with the operation of the cleaning machines. Surprisingly large objects such as baulks of timber occasionally arrive at some works.

Typically, coarse screens consist of 25 mm bars at 150 mm spacing. Even with mechanical cleaning gear, some manual attention is needed, so mechanical gear is often omitted.

Coarse screenings can be disposed of by burying. When sufficient labour is available, the smaller objects may be returned to the flow and dealt with by the main screens. In some cases it is feasible to have coarse screenings removed by the refuse collection service.

9.3 Treatment of screenings

A common method of screenings treatment is the use of a macerator or dis-integrator pump. This reduces particle size sufficiently for the solid matter to be returned to the main flow from which it is subsequently removed in the primary sedimentation tanks.

A disintegrator is a special form of centrifugal pump with cutting blades on the impeller and in the casing. Screenings are dropped by the screen-cleaning gear on to a conveyor belt or into a trough spanning the screen channel and passed to a well from which the disintegrator draws. A connection between the channel and the well provides diluent water. The disintegrator discharges to the screen channel. Where a flume is provided to convey the screenings to the well, a branch from the discharge line provides flow along the trough.

Commonly, the disintegrator is automatically controlled in association with the screen-cleaning gear. When the cleaning gear is actuated (by differential floats or time switch), the disintegrator is started. Flow is therefore passing along the screenings trough by the time material is dropped into it. The disintegrator continues to run for a few minutes after the cleaning gear has ceased to operate.

Cutting blades need to be re-edged every few months. Spare sets of blades are kept in stock so that machines need be out of service for no longer than is needed to change the blades.

Disintegration can be criticised on the grounds that screenings are so readily removed from sewage that it is illogical to return them in a form which adds to the sludge disposal problem. The burying of screenings on a remote part of the site was once common and is still practised at small works. Provided there is not too much material to dispose of, natural decay will occur in an inoffensive manner.

Incineration of screenings is another alternative. Where this has been adopted, it has often been considered advisable to wash the screenings first so as to remove fine matter and grit. The screenings are washed by jets of water (or treated effluent) on an oscillating table. Washings pass through perforations in the table and are given full treatment with the main flow.

Incineration has only rarely been adopted as there is some reluctance to add further items of plant which would need operation and maintenance. Possible air pollution, or expedients to avoid it, is another factor.

At several recently constructed plants, screenings are washed and automatically loaded into plastic sacks for removal by the refuse collection service.

9.4 Size of screens

With automatically operated, mechanical cleaning gear, mean velocity in the screen channel is the main criterion for screen size. The approach velocity may be up to 0.9 m/s without the afflux becoming excessive. If the velocity is lower than 0.3 m/s, grit will be settled in the chamber. The consequences of grit settlement are not likely to be very serious. A plug valve in the floor of the chamber enables deposits to be flushed out. In some areas large quantities of screenings sometimes enter the works during exceptional storms. Screen area needs to be generous in these areas if the cleaning gear is not to be overtaxed. Due to the presence of downstream controls, mean velocity through the screen chamber will vary with depth and discharge. This is discussed in more detail in Section 9.14.

At all but the smallest works, more than one screen chamber is provided. Where there are two channels, one may have mechanical gear while the second, used as a standby, has a hand-raked screen. Sometimes the standby screen will be of the same pattern as the mechanically raked screen so that mechanical gear can be added if the works is expanded.

On larger works, the total width will be divided between two or more working channels so that individual screens are not too large. If the work is divided between several units, standby capacity will not usually be necessary. Where there are multiple channels, penstocks will usually be provided so that some channels can be kept out of service during periods of low flow.

Widening of feed channels to form screen chambers should not be abrupt or the main flow will break away, leaving eddying flow in the corners trapping screenings and grit. An angle of splay of $12\frac{1}{2}°$ is a commonly recommended minimum for open-channel transitions.

9.5 The comminutor

This device combines screening and disintegration. It consists of a slotted sheet-brass cylinder rotating on a vertical axis through which the whole of the flow passes, Fig. 9.5. The outlet is by way of an inverted siphon pipe below the cylinder.

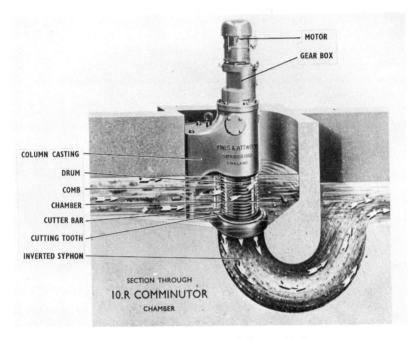

COLUMN CASTING
DRUM
COMB
CHAMBER
CUTTER BAR
CUTTING TOOTH
INVERTED SYPHON

MOTOR
GEAR BOX

SECTION THROUGH
10.R COMMINUTOR
CHAMBER

Fig. 9.5   The Comminutor (*Courtesy of Jones & Attwood Ltd*)

Solids arrested on the outside of the cylinder are shredded by fixed cutting edges as the cylinder rotates until they will pass through the slots with the flow. The comminutor avoids any handling of screenings, mechanical or otherwise, and requires very little power. For a small works which is left unattended for long periods, the comminutor is a good choice provided an electricity supply can be provided without too much expense.

Wear on the cutting edges is usually reduced by installing comminutors downstream of grit removal tanks.

The head loss through a comminutor is greater than that through screens. This may mean that subsequent treatment units will have to be at lower levels than with a screen. On some sites this would result in increased cost.

Comminutors are available in sizes up to 0.9 m diameter. Where screenings are to be removed from the flow a Screezer can be used instead of a comminutor.

9.6 Grit removal

Particles of grit (or 'detritus') are relatively easy to remove from sewage. They are settled out at an early stage of treatment, so that the lighter sludge which is removed later will be grit-free. If this were not done, the grit would accumulate at various points in the sludge system and cause excessive wear in sludge pumps.

The amount of grit in sewage varies from 25 to 250 mg/l. The concentration may be particularly high during storms, especially where the sewers are laid to flat gradients.

The specific gravity of grit particles is about 2.65 while that of organic particles is very close to 1.0. Consequently the settling properties of grit particles are of different order from those of organic particles. This makes it possible to settle out a fairly clean grit and also to wash organic matter from grit so that the final product is innocuous and can be easily disposed of by tipping or can be used for filling depressions or even as bedding for footpaths.

9.7 Sedimentation processes

In order to establish a basis for the design of tanks for the removal of grit particles from sewage by settlement it is first necessary to see how an isolated particle settles in an infinite body of fluid.

When the particle is released it will accelerate under the action of gravity, the force being

$$V(\rho_s - \rho)g$$

in which V = volume of particle,

         ρ_s = density of particle,

and ρ = density of fluid.

Counteracting the gravitational force is a force due to the drag of the fluid on the particle which may be expressed as

$$\tfrac{1}{2}C\rho Av^2$$

in which A = the projected area of the particle in the direction of motion,

         v = the velocity of the particle relative to the fluid,

and C = the coefficient of drag, which will depend on factors such as the viscosity of the fluid, the shape and roughness of the particle and the flow pattern around it.

Initially, the velocity is zero so that, at release, the full gravitation force acts alone, but as velocity increases the drag force opposing settlement will become larger until a stage is reached when the two forces balance. Settlement will then continue with uniform velocity.

This velocity, the *terminal* velocity v_s, can be obtained by equating the expressions for the gravitational and drag forces:

$$V(\rho_s - \rho)g = \tfrac{1}{2}C\rho Av_s^2 \tag{9.1}$$

In the cases considered here the acceleration phase is so short that it can be neglected and the terminal velocity is the factor in which interest lies.

This formulation applies equally to the rising velocity of a particle less dense than the surrounding fluid.

The next steps are easier to pursue if the particular case of a spherical particle is taken, for which the terminal velocity is

$$v_s = \sqrt{\frac{2g}{C} \cdot \frac{2d}{3} \cdot \frac{\rho_s - \rho}{\rho}} \tag{9.2}$$

where d = diameter of sphere.

Dimensional analysis suggests and experiment confirms that C depends on a Reynolds number expressed as $\mathscr{R} = v_s d/\nu$.

For a laminar flow around the particle $C = 24/\mathscr{R}$ and the expression reduces to

$$v_s = \frac{g}{18\nu} \cdot \frac{\rho_s - \rho}{\rho} \cdot d^2 \tag{9.3}$$

This is the well-known Stokes law for a sphere falling through an infinite viscous fluid. It applies when the Reynolds number is less than 1.0.

For larger values of the Reynolds number, an empirical relationship for C has been found:

$$C = \frac{24}{\mathscr{R}} + \frac{3}{\sqrt{\mathscr{R}}} + 0.34 \tag{9.4}$$

In this case v_s and d cannot be related explicitly but if they are regarded as the design variables and ρ_s, ρ, ν and g are taken as having fixed values, it is possible to relate dimensionless design parameters for v_s and d independently on a graph (cf. the Wallingford charts, Chapter 6).

The parameters are

$$\mathbf{V}^3 = \frac{\mathscr{R}}{C} = \tfrac{3}{4} \cdot \frac{\rho}{\rho_s - \rho} \cdot \frac{v_s^3}{g\nu} \text{, independent of } d \tag{9.5}$$

and

$$\mathbf{d}^3 = C\mathscr{R}^2 = \tfrac{4}{3} \cdot \frac{\rho_s - \rho}{\rho} \cdot \frac{gd^3}{\nu^2} \text{, independent of } v_s \tag{9.6}$$

The relationship is shown in Fig. 9.6. Grit particles have a specific gravity of about 2.65 and subsidiary dimensional scales enable v_s to be related to d when $g = 9.81$ m/s^2 and $\nu = 1.13 \times 10^{-6}$ m^2/s. In order to design a grit-settling tank it is necessary to know the settling velocity of the smallest particles which it is desired to remove and dimensional scales are provided for this purpose. Though the graph applies only to spherical particles and there are some further complications to be discussed below, it is adequate for practical purposes.

The discussion has been confined so far to the laws of settlement of isolated particles in an infinity of fluid. If particles are close to each other or to the walls of the vessel, the flow patterns around them, and consequently the value of C, will differ. Also where the particles are in high concentration the fluid will be displaced upwards.

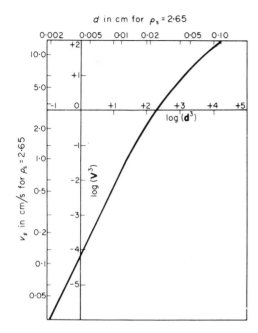

Fig. 9.6 Relationship between diameter and settling velocity

A further effect of high concentration occurs when the suspension consists of a mixture of particles with different settling velocities. The faster particles will collide with the slower ones. On the average the large particles will settle more slowly and the light particles more quickly than normal. At particularly high concentrations, the overall effect will be that of a mixture of particles all settling at much the same rate, which will be some form of average of the free settling rates of the individual particles. This is known as *zone settling* and is described in Section 10.5.

Grit-tank design is not sufficiently refined for either of these effects to be taken into account. If they were to be taken into account it would be easier to obtain settling velocity data directly from tests in quiescent columns with sampling points at several levels than to attempt to measure the factors which would be needed to obtain a result by calculation.

A concept which has had a good deal of influence on settling-tank design was offered by Hazen.[1]

This is shown in its simplest form by Fig. 9.7 which represents a longitudinal section through an idealised rectangular settling tank. Within the settling zone, fluid velocity is horizontal and uniform throughout the depth. The inlet and outlet zones are provided to ensure uniformity of velocity within the settling zone. Any particles which reach the bottom of the settling zone pass into the sludge zone and stay there.

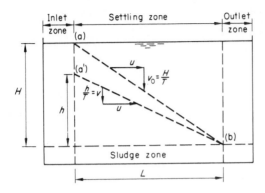

Fig. 9.7   Ideal rectangular settling tank

An element ΔB of tank width will be considered and fluid velocity will be assumed to be uniform across the element.

The velocity of flow through the settling zone is $Q/(H \cdot \Delta B) = u$ and the time of flow through the zone is $L/u = LH \cdot \Delta B/Q = T$.

A particle initially on the surface at (a) will just reach the sludge zone at (b) if its settling velocity v_0 is constant and equal to H/T which may be expressed as $Q/(L \cdot \Delta B)$ and is therefore independent of the actual values of either H or T. That is, for any given value of $Q/(L \cdot \Delta B)$, the ratio H/T will be the same. Any change in H is accompanied by a proportionate change in T.

The ratio $Q/(L \cdot \Delta B)$ is the rate of flow per unit surface area and is termed the overflow rate or surface-loading rate. Alternatively, this parameter can be described as the depth per unit of detention time.

It is evident that a tank of this ideal form will remove all particles of settling velocity equal to or greater than v_0. Particles which are lower than (a) at the inlet and which have settling velocity v_0 will reach the sludge zone before the outlet zone is reached and this will be true also for particles of greater settling velocities whether they are at or below the surface initially.

In order to obtain a result for particles settling at velocities less than v_0, it is necessary to make some assumptions as to their distribution over the inlet plane of the settling zone. The simplest assumption is that they are uniformly distributed. In this case the proportion of particles below (a'), which is h above the floor, will be h/H. Considering particles with settling velocity $v = h/T$, all those between (a) and (a') will remain in the flow passing out of the tank since they will not have time to reach the sludge zone, while all those below (a') will be removed from the flow. The proportion removed is therefore h/H which is equal to v/v_0. This result is independent of the actual values of H and T, depending only on their ratio H/T, that is upon the surface loading rate.

The Hazen law is more general than this simple case shows. It applies to idealised radial flow circular tanks whether the flow is outwards or inwards. In these cases though the fluid velocity changes in the direction of flow the

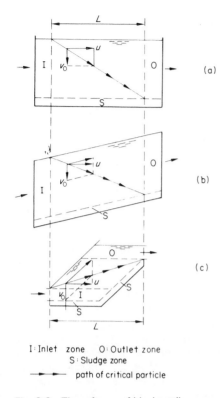

I: Inlet zone   O: Outlet zone
S: Sludge zone
——►—— path of critical particle

Fig. 9.8   Three forms of ideal settling zone

surface-loading rate remains constant. The settling particles follow paths which are curved in elevation.

The fluid velocities need not be horizontal. An ideal tank of the form shown in Fig. 9.8(b) will produce the same results as a tank in which the top and bottom are horizontal, provided that the plan area is the same.

An ideal settling zone of the type shown in Fig. 9.8(c) is of some interest. Here the inlet and outlet planes are horizontal. Detailed consideration, which will not be given, leads to the conclusion that Hazen's law still applies provided that the surface area is based on the length L. Strictly, therefore, the area used in calculating critical settling velocity should be defined as the net area traversed by the flow in the horizontal plane. In a truly vertical settling zone this area is zero and no particles with settling velocities equal to or less than the ratio of the discharge to the actual plan area will be removed. In real upward-flow tanks (Section 10.2, 3), the flow traverses the area of the tanks from the centre to the peripheral outlet weir so the flow is by no means truly vertical.

Velocities need not be uniform throughout the depth of a tank for a particle

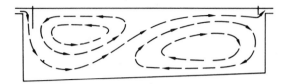

Fig. 9.9   Flow pattern in a rectangular sedimentation tank (longitudinal section)

initially at the surface to reach the floor at the same distance along the tank as if the velocity were uniform. The forward movement of the particle as it settles depends on the mean velocity of the layers through which it has travelled, independently of the way in which the velocity may have varied. Curvature of flow paths in elevation has been shown not to affect the amount of flow traversed vertically by a particle settling in a given time.[2] Even flow patterns with large eddies like those shown in Fig. 9.9 have been found to give the same degree of settlement of discrete particles as that predicted for an ideal tank.[3] However, this might be true only to the extent that there is turbulent mixing between the main flow and the eddies.

Lateral variations in velocity remain to be discussed. In considering an ideal rectangular settling zone, a small element of width was taken. If adjacent elements have different ratios of flow to width, it is evident that they will have different surface-loading rates and consequently that the removal of suspended particles will differ from one element to another. Considering particles of settling velocity equal to the mean surface-loading rate for all the elements, these will be totally removed in all those elements with surface-loading rates equal to or less than the mean, but only a proportion of these will be removed in the remaining elements. Consequently, as Fischerström[4] has shown, surface-loading rate must be uniform across the tank: slow-flowing elements do not compensate for fast-flowing elements. It can be shown that this applies to a suspension of particles with different settling velocities as well as to a suspension of particles with the critical settling velocity.

It follows that where several identical tanks are used in parallel, the flow should be divided equally between them.

Though Hazen's law is valid over a wide range of conditions, provided that surface-loading rate remains uniform laterally, it is essential for each particle to maintain a constant settling velocity relative to the fluid surrounding it. This statement needs a slight qualification, however, in that oscillation of the settling velocity of a particle about a constant mean value can be allowed. Thus turbulence of the flow and the effects of hindered and zone settling may not invalidate the results though the settling velocities will be different from those calculated for isolated particles in a quiescent fluid.

Any tendency for settling velocities to change permanently with time, such as may occur with particles which adhere on contact or flocculate, renders the

Hazen concept of little use and this is unfortunately the case with organic suspended material in sewage, though not for grit.

Though for suspensions to which Hazen's law can be applied, depth and detention time are irrelevant in theory, there is a practical limit to which they can be reduced, in that the forward velocity must not be so high that particles settling to the floor are prevented from remaining there.

9.8 Detritus settling tanks

Figure 9.10 shows an early form of detritus tank at a small works. The screen is accommodated in the tank. On a larger works, the tank would be similar to a small rectangular sedimentation tank (see Section 10.2, 1) and the screens would be upstream.

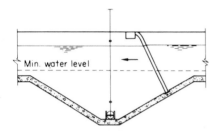

Fig. 9.10   Small detritus tank accommodating screen

These tanks, following Royal Commission recommendations, were given a capacity of $\frac{1}{50}$ DWF divided between at least two tanks. A good deal of organic material settled with the grit, especially at low flows.

The Dorr Detritor (Figs 9.11 and 9.12) and similar tanks are based more rationally on a surface-loading basis to settle grit particles equal to and greater than a chosen equivalent sieve size. Adjustable baffles across the inlet section enable a close approximation to parallel flow to be achieved. There is a low outlet weir, and the floor is only sufficiently far below the weir to accommodate the floor scraper. The weir is submerged, depth of flow in the tank depending primarily on conditions in the downstream channel which usually accommodates a venturi flume and storm weir. Allowing the depth to vary and minimising the depth of dead water below the weir are essential features in approximating to 'ideal tank' flow.

The required surface area is calculated from

$$\frac{\text{maximum rate of flow}}{\text{surface area}} = \text{settling velocity of critical particle}$$

At rates of flow below the maximum, smaller particles will be settled.

In relating settling velocity to particle size, the specific gravity of siliceous material (sand, stone) may be taken as 2.65. If it is desired to remove

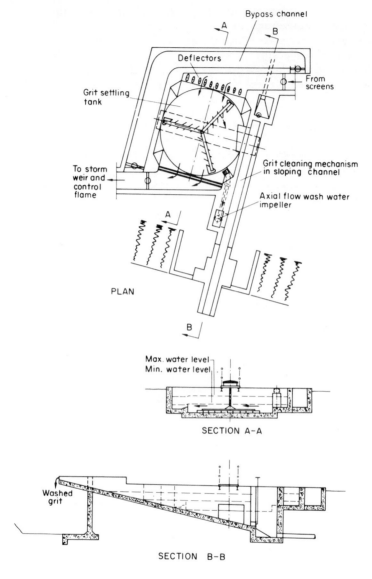

Fig. 9.11 The Dorr Detritor and grit cleaner

Fig. 9.12 Dorr Detritor at the Newton Mearns Works of Renfrew County Council (*Courtesy of the Dorr-Oliver Company Ltd*)

particles of ash or clinker, a specific gravity of 1.5 should be used. For particles of a given equivalent size this will reduce settling velocity by a factor of $(1.5 - 1)/(2.65 - 1)$. The required surface area of tank would be roughly trebled.

Design values of the settling velocity for grit range from about 15 to 30 mm/s equivalent to loadings of 54 to 108 m³/h per m² of surface.

Grit is washed free of organic material in a separate channel, described later (Section 9.11). The echelon blades of the floor scraper move settled material to the outer wall where they fall into a hopper communicating with the washing channel.

The tank is necessarily square rather than rectangular on plan on account of the use of a rotating floor scraper. The features of the downstream channel must be such that the forward velocity through the tank is not greater than about 0.36 m/s, otherwise there will be a risk of grit particles being carried through. If necessary two detritors can be used in parallel. Where there is a single detritor, a by-pass channel is always provided.

9.9 Constant-velocity channels

The size of grit particles which can be carried along by the flow in a pipe or channel is related to the mean velocity of flow. The object of the constant-velocity channel is to maintain an approximately constant mean velocity over the full range of discharge. The mean velocity is chosen so that grit particles over a given size will be deposited while the smaller particles are carried through. The use of this principle for detritus removal was first advocated by Townend[5] though it may have been used earlier.

The principle is applied by deliberately controlling the depth of flow in the channel, usually by means of a critical depth flume and adopting a shape for the channel cross-section such that the ratio of discharge to area of flow is constant.

If a flume is used the differential coefficient of discharge with respect to depth of flow, is

$$\frac{dQ}{dy} = 1.71\, b \cdot \frac{3}{2} y^{1/2}, \text{(metre second units)} \tag{9.7}$$

For the detritus channel, $Q = AU$, and for constant U,

$$\frac{dQ}{dy} = U\frac{dA}{dy} = Ut \tag{9.8}$$

in which t is the width of the water-surface at depth y.

Equating these expressions for dQ/dy,

$$t \propto y^{1/2}$$

giving the channel cross-section a parabolic form.

For a maximum discharge of Q_m, the maximum depth H and top water breadth B are chosen so that:

$$Q_m = U\tfrac{2}{3} BH \tag{9.9}$$

since the area of a parabola is two-thirds of the area of a rectangle of the same height and breadth.

A mean velocity U of 0.3 m/s is commonly aimed at. The maximum depth of the channel is often restricted by the need to avoid surcharging the incoming sewer. Consequently the top water breadth may be much larger than the depth. So that the floor will not be too flat, the theoretical cross-section is divided between several parallel channels. The parabolic shape is approximated to by plane surfaces, Fig. 9.13.

Where multiple channels are adopted, each channel is provided with its own control flume. On large works, penstocks are installed at both ends of each channel. A single channel is used for low flows, further channels being brought into use when the flow increases. The screens are placed in the inlet lengths of the detritus channels.

Grit-collecting channels are provided below the general floor level. The grit is removed by submerged pumps carried by a bridge running on rails on the side walls of the channels. The pumps discharge to a launder adjacent to the rail.

The theoretical length for a grit channel is HU/v_s in which v_s is the settling

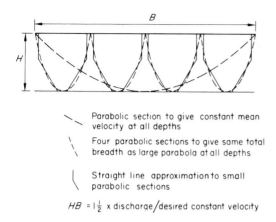

Parabolic section to give constant mean
velocity at all depths

Four parabolic sections to give same total
breadth as large parabola at all depths

Straight line approximation to small
parabolic sections

$HB = 1\frac{1}{2} \times$ discharge$/$desired constant velocity

Fig. 9.13 Cross-section of constant-velocity channels controlled by critical-depth flume

velocity of the slowest settling particle to be removed. For a parabolic channel, this leads to a surface area of $1\frac{1}{2} Q/v_s$, or 50% greater than that for an 'ideal' settling tank. Particles settling with a velocity less than v_s will in theory reach the floor but will be swept along as bed load.

Using the simple relationship $17v_s$ (Section 6.8) for the scouring velocity for a particle of settling velocity v_s, constant velocity channels should be 17 times as long as they are deep. It is usually considered prudent to make them longer than this to allow for inlet and outlet effects but opinions differ as to whether the increase should be 30%, 50% or even more. Not infrequently considerations of overall layout have an influence on the final decision.

Weir slots of the shape shown in Fig. 9.14 have been used instead of standing-wave flumes for depth control. With these the tank cross-section can be rectangular instead of parabolic. A drop is needed below the slot to secure free flow and the overall head loss will be greater than with a flume.

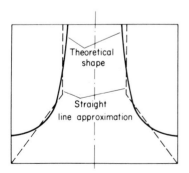

Theoretical
shape

Straight
line approximation

Fig. 9.14 Proportional weir-notch (Sutro weir) for constant mean velocity in a rectangular section channel

9.10 Aerated grit channels

These consist of tanks of substantially rectangular shape. The cross-section, Fig. 9.15, is quite large so that the mean forward velocity is considerably lower than that for other forms of grit tank. Organic matter is kept in suspension by imparting a rolling motion to the flow. This is done by injecting air as shown. In addition, incoming flow may be directed laterally, near the surface at the inlet end. The rolling motion must be sufficient to keep grit in suspension.

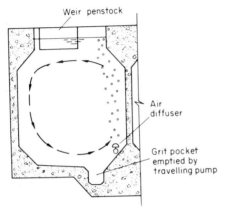

Fig. 9.15   Cross-section of aerated grit channel

Design criteria do not appear to be well established but it has been suggested[6] that the transverse surface velocity should be between 0.45 and 0.6 m/s. The size of particle retained can be adjusted by altering the air supply.

At Singapore,[7] four channels 18.3 m × 5.5 m × 3.66 m deep treat 4.24 m³/s. The air supply is 0.165 m³/s. There are installations in the United Kingdom at Derby and at Skelmersdale.

9.11 Grit-washing equipment

Even if a constant-velocity or aerated-grit channel is used, the detritus will be contaminated to some extent with putrescible matter. Disposal problems are eased by washing the grit.

The principle of the vortex or cyclone grit separator is shown in Fig. 9.16. Grit-laden flow enters tangentially at the top. The circulation of the flow, particularly in the conical lower part, causes the heavy grit particles to migrate to the wall. Washed grit is withdrawn from the apex. The greater part of the flow, carrying the organic material, is siphoned from the centre of

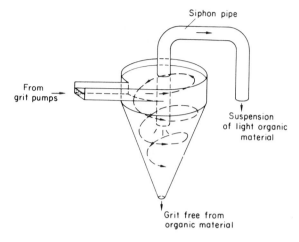

Fig. 9.16 Principle of vortex grit separator

the chamber. This is returned for treatment to the main flow through the works.

There is a compact version of this device with a horizontal axis made by the Dorr Company and known as a DorrClone. Grit-laden water is forced in through a tapered tangential inlet at the wide end. Flow carrying washed grit leaves at the narrow end while a much larger flow of liquid carrying the organics leaves axially at the wide end.

The Dorr Detritor is normally equipped with a washing channel as shown in Fig. 9.11. This device was developed originally for particle classification in the ore-mining industry. The channel slopes so that its upper part is clear of the water level which is virtually the same as that in the detritor. A reciprocating framework in the channel carries a series of combs which are transverse to the channel. The framework is driven by an electric motor. Eccentrics and swinging links cause the combs to move upwards near to the channel bed and to return at a slightly higher level. The pitch of the combs is somewhat less than the length of stroke. The motion has the effect of gradually passing grit up the slope and washing it free of organic material. To prevent the accumulation of organics in the washing channel, a propeller pump forces liquor back into the detritor, causing a circulation of liquor through the system.

Washed grit spills from the end of the channel and is received by a dump-truck or narrow-gauge-rail tipping wagon.

Where a detritor is some way below ground level it may be better to install the classifier at a higher level. It is then fed by a grit pump drawing from a small sump formed by extending the floor pocket of the detritor.

Where grit has to be pumped up to ground level a vortex or cyclone washer may be preferable to an inclined-channel classifier.

Another form of grit washer employs an Archimedean screw rotating in an inclined tube. Grit is lifted by the screw from the sump. The screw extends above water level. As the grit is worked up the tube by the screw, organic material is left behind in the water.

9.12　Separation of high flows

The main line (i.e. primary sedimentation and biological oxidation) of a sewage works serving a combined sewerage system is designed to take flows up to Q_f which is normally $3(D + E) + I$ as defined on p. 311. Maximum inflow to the works, Q_m, is two or more times Q_f depending on the setting of storm overflows on the sewerage system (see p. 84 and p. 311). Surplus flow above Q_f is diverted to the storm tanks.

Since storm tanks act as primary sedimentation tanks preliminary treatment (screening and grit removal) must be given to the sewage diverted to them.

In most cases the screens and detritus tanks are designed to deal with flows up to the maximum received, and the storm overflow weirs are placed downstream of the detritus tanks. Typical layouts are shown in Fig. 9.17. The critical-depth flume controls depth at the side weirs and renders the depth independent of controls downstream of the flume.

In order to ensure that the flow passing onwards for biological treatment never exceeds Q_f, a motor-driven penstock is linked for automatic control to the float registering approach depth at the flume. Whenever the flow reaches Q_f the penstock begins to close but is reversed as soon as the through-flow is slightly below Q_f. At all incoming flows in excess of Q_f, the penstock 'hunts' to trim the through-flow to a rate which varies only very slightly from Q_f. Surplus flow has no other course than to spill over the weirs.

While the standing-wave flume is the best form of depth-control in most cases, an orifice, module pipe or step in the channel may be worth considering on a small plant. Whatever form of control is adopted, downstream features must be designed to avoid drowning the control.

Hunting penstocks may be unnecessary if a very long overflow weir can be used. The weir has to be sufficiently long to keep the excess through-flow small at peak spill. When this is done, difficulties may arise in designing the units upstream of the weir as depth of flow will increase only slightly as the flow increases from Q_f to Q_m. The range of velocity is thus doubled.

9.13　Storm tanks

Many storms are insufficiently long or intense to completely fill the storm tanks. In these storms the tanks serve to store sewage until the storm is over. During the period of storage much suspended matter settles out. After the storm, the liquor is withdrawn through decanting arms and pumped into the main flow for full treatment.

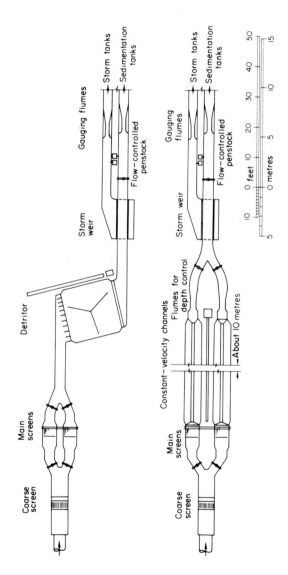

Fig. 9.17 Alternative schemes for preliminary treatment

In severe storms, the tanks fill completely and overflow to the river. They are designed to act as continuous flow sedimentation tanks so that the discharge will have received primary treatment. Mainly for this reason, the total storm-tank capacity is usually equal to 2 hours' detention at the maximum rate of flow. The frequency with which storm tanks overflow is less for large systems than for small ones on account of the storm-storage capacity in the sewers. It is arguable that relatively more storm tankage should be provided at small works to reduce the frequency of overflow.

Ideally a set of storm tanks should be arranged so that each tank is filled in turn and so that the operator has the option of preventing overflow from the first tank on the occasions when the first part of the storm flow is highly polluted. Using the first tank solely for storage has the disadvantage of reducing the volume used for continuous-flow sedimentation treatment.

Since optional use of the first tank for storage is difficult to automate, the first tank is sometimes not provided with any outlet to the river. Sometimes as a compromise a weir is provided in the dividing wall so that the first tank can overflow into the second. This principle can of course be used for all the tanks, each overflowing to the next as it becomes full, but some sedimentation performance will be sacrificed unless it is possible to arrange for the tanks to be fed in parallel once they are all full.

A great many alternative arrangements are possible for feeding storm tanks and the final choice must depend on the feasibility of using automatically operated penstocks and on the levels, in each case. Automation can often be avoided by placing inlet and outlet weirs at levels designed to ensure that the tanks fill in turn from a common channel but ultimately operate in parallel. Isolating penstocks should be provided, especially on the outlet to the first tank, so that there is some opportunity for flexibility of operation.

Most storm tanks are closely similar to rectangular sedimentation tanks (Section 10.2, 1). Small tanks are de-sludged manually after emptying. On tanks large enough to need mechanical scrapers, a single machine may be used for several tanks. The machine is moved from one tank to another by means of a transfer carriage running on rails at the outlet end of the tanks.

Provision should be made for draining feed channels into the tanks or to the tank-emptying pumping station.

9.14 Overall design of preliminary treatment plant

Alternative schemes for preliminary treatment are shown in Fig. 9.17. In the upper drawing, depths of flow in the grit tank and in the screen chambers will be controlled by the gauging flume and overflow weir. The depth–discharge relationship for the channel immediately upstream of the weir will take the form shown in Fig. 9.18. In the figure Q_m is the maximum rate of flow arriving at the works and Q_f is the maximum rate of flow to be given full treatment. Corresponding depths of flow are H_m and H_f respectively.

A similar depth–discharge relationship will apply to the screen chamber,

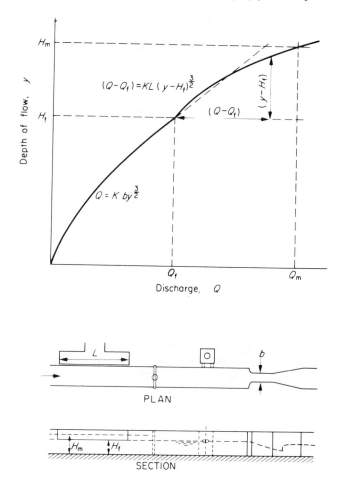

Fig. 9.18 Depth–discharge relationship upstream of storm weir

modified only slightly by the effects of the submerged outlet weir of the detritor and by the inlet vanes.

The maximum depth H_m will usually be made of the same order as the diameter of the incoming sewer, but may be modified by the need to proportion the area of the detritor to a critical settling velocity and its depth to keep through-velocity below about 0.36 m/s. The required maximum depth will depend on whether there is a single grit tank or a pair of grit tanks in parallel.

The outcome of these considerations will fix the maximum depth at the screens. The total width of the screen chambers will then depend on the velocity to be aimed at. Ideally this would be of the order of 0.3 to 0.6 m/s. However, due to the curvature of the depth–discharge relationship, mean velocity will vary with rate of flow. The relationship can be made more linear by increasing b and reducing L (Fig. 9.18). An increase in b will necessitate subsequent treatment units being placed at a lower level in order to maintain an undrowned outlet for the flume. Reducing L may result in an excessive depth of flow over the weir. Consequently it is rarely possible to produce a nearly linear depth–discharge relationship. Often H_f is a good deal closer to H_m than in the case shown in Fig. 9.18. The result of this is that a compromise has to be made, which usually involves the acceptance of quite low velocities at low flows. Where a single grit tank is used, a by-pass channel is usually provided. This has not been shown on the figure.

In the lower drawing of Fig. 9.17 each constant-velocity channel has its own control flume which governs depth at the screen chamber as well as in the detritus channel. Since the flow passes through both a control flume and a gauging flume, the total fall is likely to be somewhat greater than in the other scheme. Conceivably, the control flumes could be eliminated by suitable choice of gauging-flume breadth, overflow weir length and detritus channel cross-section. This would be satisfactory if there were only one detritus channel, or if the two channels were to work continuously.

On a large plant there would be several detritus channels, automatically called into operation as the flow increased and each would need a control flume.

An alternative to the layout shown would be to separate storm flow in advance of grit removal, different detritus channels being provided for main and storm flows. Screening would be carried out before separation in order to avoid trouble with float wells or other control gear. This arrangement has the disadvantage that there is less flexibility in the use of channels when repairs and maintenance are needed.

It is sometimes advantageous to have the floor of the inlet channel some inches below the invert of the sewer outfall. This avoids backing up the flow in the sewer at low flows (note the steep rise of the curve in Fig. 9.18 from the origin) and enables channels of greater depth and smaller width to be used. A smooth transition from the sewer to the channel is of course preferable to a step.

References

1 Hazen, Allen, On sedimentation, *Trans. Am. Soc. Civ. Engrs*, **53**, 45–71, 1904 (reprinted in *J. Inst. Sew. Purif.*, Pt 6, 521–531, 1961).
2 Fitch, E.B., Flow path effect on sedimentation, *Sew. Ind. Wastes*, **28**, 1–9, 1956.
3 Thompson, D.M., *Scaling Laws for Rectangular Sedimentation Tanks*, Ph.D. thesis, University of Manchester, 1967.
4 Fischerström, C.N.H., Sedimentation in rectangular basins, *Proc. Am. Soc. Civ. Engrs*, **81**, (687), 1–29, May 1955.
5 Townend, C.B., The elimination of the detritus dump, *J. Inst. Sew. Purif.*, Pt 2, 58–87, 1937.

6 Joint Cttee Am. Soc. Civ. Engrs and Water Pollution Control Fed., *Sewage Treatment Plant Design*, Am. Soc. Civ. Engrs, New York, 1959.

7 Oakley, H.R. and Cripps, T., The Ulu Pandan sewage disposal scheme, Singapore, *Proc. Inst. Civ. Engrs*, **21**, 13–36, Jan. 1962.

10

Sedimentation tanks

10.1 Introduction

The separation of solid particles from flowing liquor by settlement in near quiescent conditions is used in primary treatment, in the removal of humus from filter effluents and in the recovery of activated sludge. Several different forms of tank can be used for any of these purposes with minor differences.

The forms of tank will be described first and their applications to different stages of treatment will then be discussed.

Aspects of the sedimentation process which apply to suspensions of discrete particles have already been discussed (Section 9.7). Further aspects which apply to sewage are dealt with at the end of the present chapter together with such design criteria as are at present in use.

10.2 Forms of sedimentation tank

(1) Rectangular tanks

This form of tank is shown in Fig. 10.1. The length is usually $2\frac{1}{2}$ to 3 times the breadth though both smaller and larger length/breadth ratios have been

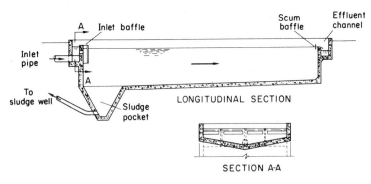

Fig. 10.1 Rectangular sedimentation tank

used. The precise choice depends primarily on considerations of site layout and structural economy. The depth is of the order of 3 to 3.6 m but again might be smaller (2.5 m or even 2 m) or larger (up to about 5 m). Since all but the smallest tanks are equipped with mechanical scrapers, the breadth is governed by the availability and cost of machines. Tanks up to 24 m wide are not uncommon.

Clearly a good deal of work could be done to optimise the dimensions of a battery of tanks in each individual case but it is probably true to say that this is done intuitively more often than by detailed analysis.

Flow moves along the tank at a mean velocity of the order of 5 mm/s but reversals of flow and large eddies have frequently been observed so that local velocities are much larger.

Inlet arrangements of a wide variety of forms have been adopted. The object is to deflect and to attempt to diffuse the incoming flow so that jets of high-velocity flow are avoided. The arrangement shown in Fig. 10.1 is not based on any particular design and is given merely to exemplify some of the ideas which have been commonly adopted. Clements[1] has shown that fairly deep baffles may be of benefit.

The outlet consists of a weir across the whole width of the tank. Since light material separates by flotation and forms a scum on the surface, a dip-plate is provided to prevent the scum from flowing over the weir. The proportions of the tank are usually chosen so that the rate of flow per unit length of weir is limited. If necessary, a double-sided trough weir can be used.

Since a large proportion of the suspended matter settles near the inlet end it is from this end that the sludge is removed. The floor is swept by scraper blades supported from a moving bridge spanning the breadth of the tank. The bridge is carried on flanged wheels which run on rails on top of the side walls. Cantilever bridges serving a pair of tanks and supported on the dividing wall are an alternative to bridges spanning individual tanks. The problems of avoiding 'crabbing' motion are somewhat reduced with a canti-lever bridge. The driving motor is most commonly on the bridge, power being supplied by a trailing cable wound on a reel. The floor is swept with the bridge travelling towards the inlet end, scrapers being raised for the return journey. Skimming blades move the scum to the inlet end. A flight of trans-verse blades is used to move the scum to one side where it may be decanted off by some form of moveable weir and discharged to the sludge well.

At the inlet end of the tank there is a pocket in which sludge can be left to consolidate and which provides some balancing storage. Sludge is with-drawn under the head of water in the tank by opening the valve in a sludge well. A row of pockets of inverted pyramid form is sometimes used instead of a single transverse pocket.

An alternative to the bridge (or 'Mieder') scraping machine is the flight scraper in which a number of scraper blades are attached to endless chains running on sprocket wheels. Scrapers of this type are sometimes used for storm tanks and are very common in the U.S.A.

Rectangular tanks at small works are de-sludged manually after emptying the tank. The floor is sloped to the inlet end but no sludge pocket is provided. Sludge leaves the tank by plug valves in the tank floor and flows by gravity to a separate sludge well from which it is pumped to drying beds.

(2) Circular tanks

The circular tanks shown in Fig. 10.2 and Fig. 10.5, are typical of designs used for primary sedimentation. Depths are about the same as for rectangular tanks. Diameters most frequently range between 15 and 30 m though both smaller and larger tanks have been constructed. The feed pipe passes under the tank and rises vertically at the centre, terminating in a bell-mouth. A cylindrical baffle plate prevents flow from streaming across the surface. Flow leaves the tank over a peripheral weir protected by a dip-plate.

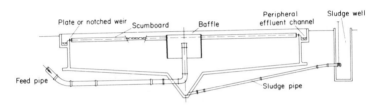

Fig. 10.2 Cross-section of a radial-flow circular sedimentation tank. The scraping machinery and centre supports are omitted

   The flow is intended to be radial and nearly horizontal, the velocity decreasing outwards. As with rectangular tanks, complex flow patterns have been observed. Hubbell[2] introduced a number of modifications to the inlet pipe and centre baffle which produced more uniform flow patterns. Townend and Wilkinson[3] suggested that there might be some advantage in making the profile of the floor a hyperboloid of revolution such that the cross-sectional area of flow (a cylindrical surface) was constant at all radii, and tanks of this form have been constructed.
   The floors of circular tanks are most commonly swept by blades supported from a rotating radial bridge. The wheels running on the outer walls may be rubber-tyred, or flanged wheels running on a rail may be used. With rubber-tyred wheels, force along the bridge has to be taken by the supports of the centre platform. Rails have been known to creep round the tank and to loosen the holding-down bolts.
   Single scraping blades of spiral form have been used but multiple blades in echelon at an angle to the direction of motion are more common. Scum is typically collected by a skimming blade under the bridge and moved outwards by a flight of transverse blades, but detailed arrangements vary with the manufacturer.
   The conical central pocket shown in Fig. 10.2 is for sludge collection and consolidation. Sludge is drawn off under the pressure of the tank contents.

Circular tanks for humus or activated-sludge settlement are of smaller diameter than those for primary treatment and have a floor slope of about 30° to the horizontal. Tanks of this form should perhaps be described as 'semi-upward flow' tanks. A central pocket is not provided as the sludge has to be removed quickly and continuously. Though bridge-driven scrapers have been used for tanks of this form, ring-driven scrapers are an alternative. The scraper is supported and moved by a steel-plate ring running round the tank just inside the weir driven by a stationary motor.

Flat-floored tanks with Vee scrapers (see Fig. 10.3) and siphons or pneumatic pumps for sludge collection are a good choice for activated-sludge settlement as they can be of larger unit size, effecting some saving in structural cost.

Fig. 10.3 Final settling tank with vee scrapers (*Ames Crosta Babcock Ltd*)

Scrapers driven from a fixed bridge and supported from beneath by a foot-step bearing are commonly used for thickeners in the coal and ore industries but are rare in sewage-treatment practice.

Walton and Key[4] developed a spiral-flow circular tank for water treatment at Alexandria. Flow entered tangentially at the periphery and left over a weir forming a sector of the periphery. Spiral flow was encouraged by the presence of a column in the centre of the tank. Small-scale tests suggested that performance would be better than with a centre-feed radial-flow tank of the same proportions. The strength of the circulating motion rendered the flow less likely to be affected by wind. The full-size tanks (33.5 m in diameter) operated successfully.

Vokes and Jenkins[5] tried out this idea at Birmingham for sewage treatment, adapting one tank for spiral flow and operating it in parallel with a centre-feed tank. The spiral-flow tank did not give improved performance but it may be noteworthy that the tanks had quite different proportions from those of Walton and Key.

(3) Upward-flow tanks

These are most commonly square on plan as shown in Fig. 10.4, though a few circular tanks exist and this was probably the original form. The upper part of the tank has vertical walls while the lower part is of inverted pyramid form with sides sloping at about 60° to avoid the adherence of sludge. Flow enters through the feed pipe which terminates in an upward-facing bellmouth. A square or cylindrical baffle prevents flow from passing across the surface. Emerging from beneath the baffle, the flow passes upwards in the zone between the baffle and the walls, leaving the tank over a weir around the tank. The baffle box (and sometimes the feed pipe) is supported by beams across the tank.

The depth of the baffle and of the inlet pipe vary considerably in different designs and there seems to be no agreed opinion as to the best arrangement. Though surface-loading rate has been the usual criterion for the design of upward-flow tanks, detention period is probably important and consequently a fairly deep baffle may be advantageous. On the other hand, the lower part of the tank is provided for sludge storage and the baffle must not be so deep that settled sludge is disturbed.

Sludge is drawn off under the pressure of the tank contents. Neither power-driven machinery nor manual labour is required, giving upward-flow tanks advantages over other forms for small installations.

The size of an upward-flow tank is limited by the depth of excavation occasioned by its pyramidal shape and ranges from 4.5 to 9 m square. As unit size is limited, upward-flow tanks are more expensive than other forms for large installations. Larger tanks with the floor divided into four pockets and with four feed points over the centres of the pockets together with trough weirs across the tank would be a possibility. This form has been used for water treatment but is likely to be more expensive than a rectangular or circular tank of similar size. In water treatment, flow is introduced at the bottom of the tank and moves upwards through the sludge, encouraging flocculation. Upward-flow tanks are frequently preferred in water treatment for this reason.

Since the weir length of an upward-flow tank is large relative to the surface area (and hence relative to the flow), patchy flow may occur over the weir if it is not quite level. The risk of patchy flow increases with decreasing size of

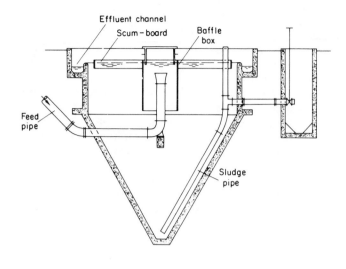

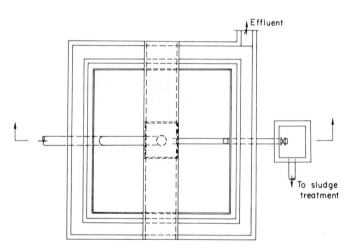

Fig. 10.4 Upward-flow sedimentation tank

tank. This problem may be overcome by using a steel-plate weir, with V-notches instead of a continuous sill.

10.3 Application of the several forms of sedimentation tank

(1) Primary sedimentation

There is no conclusive evidence that one form of tank is markedly superior in performance to another. Consequently the total volume and total surface area of the tanks provided for a given scheme would be much the same whether rectangular, circular or upward-flow tanks were used. Primarily, constructional cost is the guiding factor and may be related to the size of the plant and the configuration of the site.

The total capacity is almost always divided between at least two units so that there will be a measure of flexibility in operation and some capacity will remain available when maintenance has to be carried out. Four or more tanks are often provided even where fewer larger tanks could be used.

Upward-flow tanks tend to be the most costly per unit of volume of surface area. However, sludge can be removed from them without the use of either manual labour or power-driven machinery. Consequently for the smallest works, upward-flow tanks can be a good choice.

Where a power supply can be laid on to a small works, circular tanks with scraping machines will be worth considering, especially if the site is such that deep excavation would be difficult.

Many of the older small plants have rectangular tanks without scraping machines. For de-sludging, a tank is emptied of liquor and the sludge is moved to an outlet valve manually, using a long-handled squeegee. The moisture content of the sludge, and therefore the bulk to be disposed of, may be less than that of sludge from a tank which is kept full but labour is an increasing problem. Since tanks of this type are de-sludged less frequently and have to be emptied for de-sludging, it has been usual to allow detention periods in design up to twice that for the mechanically de-sludged tanks.

Upward-flow tanks are out of the question for all but small works so elsewhere the choice lies between rectangular and circular tanks. A circular tank of the same volume and surface area as a single rectangular tank will usually cost less to build, as the area of wall is somewhat less and the walls themselves can often be slightly thinner since the cantilever moments at the base are relieved by ring action to a greater extent than those of straight walls are relieved by beam action. The importance of these factors depends on the size of the tank.

Common division walls may be used for a battery of rectangular tanks so that for a group of tanks the rectangular form may be cheaper to construct. Furthermore, the gross site area occupied by a group of rectangular tanks will be less and there will be no intervening spaces to be paved or turfed.

Though rectangular tanks have these advantages, circular tanks have been very frequently chosen for large works. One factor is the simplicity and neatness of radial scraping machines.

Primary tanks clarify sewage by flotation as well as by sedimentation. Suspended matter less dense than the fluid, such as oily, greasy or fatty material, will rise to the surface to form a scum. This is retained in the tank by dip-plates protecting the weir. It is moved along the surface by skimming blades carried from the scraper bridge to a decanting device for discharge to the sludge well as illustrated in Fig. 10.5. Decanting devices differ in detail but several share the principle of a horizontal pipe with a slot along its length.

Fig. 10.5   Primary settling tank: scum removal device *(Ames Crosta Babcock Ltd)*

The pipe is attached to pipes leading to the sludge well by a rotating joint. When the skimming machinery is operating, an arm attached to the slotted pipe engages with the bridge in such a way that the pipe is rotated about its axis until the lip of the slot is just below water level. Scum then flows into the pipe and passes to the sludge well.

Whenever a suspension containing particles of different density from the fluid is passed through a tank at low velocity, clarification by both settlement and flotation will occur. Fundamentally, either the settling or the flotation process will be the design criterion. Thus a tank to treat a liquor with a good deal of low-density material, which rises more slowly than the heavy material falls, should be designed as a flotation tank rather than as a sedimentation tank. However, primary tanks at sewage works seem to fall into the opposite category, where settlement is the critical process, and flotation is not taken into account in deciding the dimensions of the tank.

Where a sewage or trade waste contains a good deal of grease, it is usually considered advisable to treat the grease separately from the sludge. Indeed 'recovery' is a more accurate description than treatment as the grease may be subjected to refining processes and sold.

If grease and similar floating material is to be deliberately recovered, it may be advisable to use a flotation tank in advance of primary sedimentation. A rising current of air bubbles from diffusers on the tank floor (Section 11.8) will assist flotation and restrict sedimentation. Flotation tanks do not need to be so large as sedimentation tanks. Though the total tankage will be greater if separate flotation tanks are used, there may be an advantage in isolating the flotation process.

Due to the presence of grease in sewage, trouble is sometimes experienced where the form of a sedimentation tank is such that there are horizontal surfaces with liquor beneath them. Many inlet arrangements involve details of this kind. A layer of grease accumulates on the underside and must be removed periodically after emptying the tank. Severe trouble of this nature is not common but, where a sewage is expected to contain more grease than usual, downward-facing submerged horizontal surfaces should be avoided as far as possible.

(2) Storm tanks

As the use of storm tanks is intermittent, a battery of rectangular tanks can be served by a single scraper machine. A transfer carriage is used to move the machine between tanks. The rectangular form lends itself to arranging for tanks to be filled in turn, possibly by overflowing from one to another over the division walls. For these reasons, rectangular tanks are the almost invariable choice for storm tanks at sewage works. However, when works are remodelled and extended it is not uncommon to use the old sedimentation tanks of whatever form as storm tanks.

The off-sewer storm tanks at Coventry[6] are circular. These tanks give only a short detention at high rates of flow and it was considered advisable to operate scrapers continuously while the tanks were in use to minimise the amount of settled sludge in the tank. Furthermore, there is an advantage in passing sludge to the sewer at not too high a rate. As the tanks operate automatically circular tanks with rotary scrapers had much to recommend them.

(3) Humus tanks

The capacity of a set of humus tanks is smaller than that needed for primary sedimentation. Consequently, upward-flow tanks are often used for humus settlement even on works where rectangular or circular tanks are used for primary treatment. It used to be customary to give humus tanks at single-pass filter plants about half the capacity of the primary tanks but more generous provision has sometimes been made more recently.

Semi-upward-flow circular tanks are commonly used on the larger plants.

(4) Activated-sludge settling tanks

It is imperative to return settled activated sludge as quickly as possible to the aeration tanks. As circular tanks lend themselves best to continuous sludge removal they are the almost invariable choice.

A number of plants have semi-upward-flow tanks of about 9 to 15 m diameter with floor slopes of about 30°. Ring-driven scrapers are commonly employed in tanks of these proportions.

It is possible to use tanks of large diameter with flat doors, if suitable scraper machines are installed. Vee scrapers are available driven by a rotating radial bridge. Sludge collecting at the apex of the Vee is drawn off by siphon pipes or raised by 'Pneu' pumps. Side-wall scraping blades are sometimes provided for these tanks, Fig. 10.3.

The capacity of activated-sludge settling tanks has often been based on a detention period of $1\frac{1}{2}$ to 2 hours at the maximum rate of flow. Since the maximum rate of flow may be up to 4 × DWF (3 × DWF throughout plus 1 × DWF re-cycled activated sludge) the equivalent DWF detention period is 6 to 8 hours, making total final tank capacity of the same order as primary tank capacity. Limiting the flow per unit length of weir (see p. 228) is probably more important for final tanks than for primary tanks.

The mass-flux theory (Section 10.5) is now coming into use as a more rational approach to the design of activated-sludge settling tanks.

10.4 Primary settlement

The process which takes place in sedimentation tanks may be studied by means of tests of the suspension in a settling column, Fig. 10.6. Results typical of sewage at the stage of primary treatment are shown in Fig. 10.7 in which it can be seen that the concentration of suspended solids decreases with time and at a decreasing rate. Over most of the depth of the column the

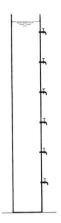

Fig. 10.6   Settling column

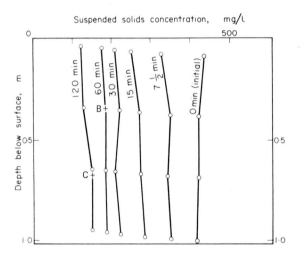

Fig. 10.7   Typical behaviour of primary tank influent in quiescent settling column test (results are extremely variable)

concentration is fairly uniform at a given time. An important feature not apparent from a single set of results like those of the figure is that the extent of clarification in a given time is strongly dependent on initial concentration.

An idealised view of sedimentation tank treatment can be obtained by assuming that the suspension passes slowly through the tank as though it were within a test column, clarifying as it moved along. At the outlet end, concentration would be similar to that in a test column after a period equal to the time of flow, or detention time, of suspension in the tank. The concentration of suspended solids in the effluent could be assumed to be the same as the average concentration throughout the depth of the test column measured at a time equal to the tank detention time.

Suspensions of heavier, non-flocculent particles in low concentration behave in a different manner to that shown in Fig. 10.7, giving results like those shown in Fig. 10.8. In these suspensions of discrete, or hydrophilic, particles, the sedimentation process can be straightforwardly related to the settling velocities of the particles, and test column results can be used to obtain frequency distributions for the settling velocities of the particles initially in suspension. Thus, taking a point such as A on the graph it can be deduced that a proportion c_a/c_o of the particles has settling velocity equal to or less than h_a/t_1 since at time t_1, a concentration c_a remains in suspension at a depth of h_a and has not had time to settle below this level. From a number of selected points a graph showing concentration against settling velocity equal to or less than a given value can be constructed.

Considering a point such as B on Fig. 10.8, and assuming that the suspension moved through a tank of depth h_b in detention time t_1 as though it were in a test column, the effluent would have a concentration equal to the mean of the t_1 curve down to B.

With a suspension in which the settling velocity of each particle remained constant, curves for successive times would have similar shapes, giving the same settling velocities, differing only in that the depth at a given concentration would increase in proportion to the time. Thus if t_2 is equal to $2t_1$ in the figure, h_2 would be equal to $2h_b$. It can be seen therefore that a tank of depth h_2

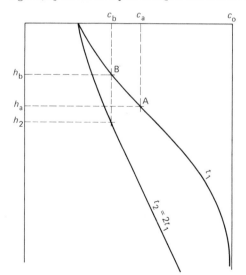

Fig. 10.8 Settling column results for a suspension of discrete particles in low concentration

with detention time t_2 might be expected to give the same quality of effluent as a tank of depth h_b with detention time t_1. What these tanks have in common is the same value for the ratio of depth to detention time, h/T, that is the same surface loading rate of flow divided by plan area, since T is equal to volume divided by flow. This is another view of Hazen's Law, Section 9.7.

Looking back at Fig. 10.7 and comparing points B and C which correspond to the same surface loading rate, it can be seen that, in contrast to the suspension in Fig. 10.8, they do not give the same degree of treatment which in this case is dependent on time. Thus Hazen's Law is of little relevance to the mode of settlement occurring in primary tanks.

Settling column tests have been found to give a reasonably good indication of the performance of tanks,[7] effluent quality being approximated by the mean concentration in the column test over a depth equal to that of the tank at a time equal to the detention period.

Before leaving settling behaviour as revealed by column tests, there is a third variety which is of interest. Here the particles are flocculent, adhering to each other on contact. The larger particles thus formed settle more quickly so the settling velocities in the suspension tend to increase with time, an effect which is greater when the initial concentration is higher. The settling column results resemble those of Fig. 10.8 at first sight but when curves for successive times are compared it is found that the later one is lower than would be expected for a non-flocculent suspension, indicating that settling velocities have been increasing during the intervening period. With suspensions of this kind, both surface loading and detention time have significant effects on predicted performance of tanks.

For the design of primary sedimentation tanks at sewage treatment works both surface loading and detention period have been used as the basis of standards. Where a surface loading criterion is used it is divided into the flow rate to give the required surface area of the tanks, leaving the depth to be decided on other, perhaps economic, grounds. A specified detention period enables the volume of the tanks to be calculated, again leaving depth to be decided separately. Of course a given depth associated with a given surface loading rate implies a certain volume and hence detention period. Correspondingly a combination of depth and detention period determines the surface loading rate. It is the inter-relationship of these factors which enables observed performance to be interpreted according to either the surface loading or the detention period criterion, and hence to each of the criteria having its advocates.

In designing a conventional plant, pragmatic designers have become accustomed to using values of both parameters which have been found by experience to be satisfactory and which in combination lead to a mean depth which intuitively seems economic. Typical values, used with maximum flow, have been detention period 2 to $2\frac{1}{2}$ hours with surface loading of 1.2 m/hr (sometimes expressed as 29 m³/hr per square metre of tank surface). Mean depth is given by the product of the parameters, thus $2\frac{1}{2}$ hr times 1.2 m/hr makes the depth 3 m.

The total capacity is then divided among 2, 4, 8 or more tanks of between about 250 and 2500 m^3 capacity each, depending on the magnitude of the flow.

The daily quantity of sludge deposited in the tanks can be obtained from:

$$Q(s_1 - s_2) \text{ kg/day}$$

in which Q Ml/day is the mean daily flow and s_1 and s_2 are the suspended solids concentrations in mg/litre of the inflow and outflow respectively. The expression gives the dry solids weight but the sludge withdrawn from the tanks has a moisture content of 95 % or more. The higher moisture contents occur where consolidation capacity of the tanks is limited and where higher proportions of returned humus or surplus activated sludge, particularly the latter, are present.

The daily quantity of BOD passing on to secondary treatment is Qb_2 in which b_2 is the BOD concentration of the outflow. If data on BOD in solution and BOD per unit SS (suspended solids concentration) are available for the particular sewage, a more refined estimate of the outflow BOD can be obtained by assuming that only the BOD associated wtih suspended solids is affected by the process.

With tanks giving 2 to $2\frac{1}{2}$ hours of detention to maximum flow, and about 6 hours to average flow, suspended solids concentration (SS) is reduced by about two-thirds and BOD by about one-third, though both higher and lower fractions are not uncommon. For a mean flow of 10 Ml/d with input BOD and SS of 330 and 300 mg/l respectively, the daily sludge production would be expected to be:

$$10 (300 - 100) = 2000 \text{ kg/d}$$

and the daily BOD passing to secondary treatment

$$10 \times 220 = 2200 \text{ kg/d}$$

With a moisture content of 96 % by weight the daily quantity of wet sludge would be

$$2000 \times 100/(100 - 96) = 50\ 000 \text{ kg/d}$$

The density of the wet sludge is given by $\rho \times 100/(m + (100 - m)/r)$ in which ρ is the density of water (1000 kg/m^3), m is the percent of water by weight and r the relative density of sludge solids often taken as 1.4. For $m = 96\%$, the density is thus 1012 kg/m^3 and the daily volume 50 000/1012 = 49.4 m^3.

If a smaller provision of sedimentation capacity is to be considered, a relationship is needed between detention period, or surface loading, and effluent concentration. As with column tests, influent concentration is highly significant. Empirically the relationship can be expressed as

$$s_2 = kT^{-a}s_1^b, \text{ but } s_2 \nleqslant s_1$$

in which T is the detention period. As the constants k, a and b can be obtained

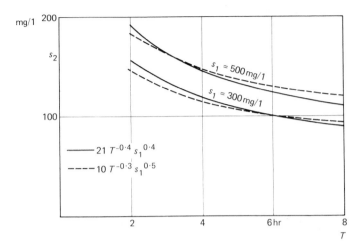

Fig. 10.9   Curves to show the effect of the values of the exponents in $kT^{-a}s^b$. Values of the coefficient k have been chosen to make s_2 one third of s_1 for $T = 6$ hr and $s_1 = 300$ mg/l

only by the analysis of observed performance, the data are mostly in the conventional range of operation, so such formulae need to be treated with some reserve. In the WRC optimising model[8] (using surface loading, however) both a and b are equal to 0.4 and k can be varied according to the expected settleability of the sewage. Evidence for making the exponents other than equal is not strong but suggests[7] that b might be larger and a smaller with appropriate change in k. For $s_2 = 100$ when $s_1 = 300$ mg/l with T as 6 hr and a and b both as 0.4 the value of k would be 21. This relationship is shown in Fig. 10.9.

Commonly, the inflow to primary sedimentation tanks contains, intermittently, humus or surplus activated sludge returned from the oxidation stage with the object of having the secondary sludge mix with the primary sludge. These secondary sludges settle fairly readily and a common assumption, with some observational support, is that there is no need to modify s_1 to account for their presence, the implication being that they will re-settle in total in the primary tank without affecting the settlement of primary suspended solids. If s_1 is modified to allow for returned secondary sludge before applying a formula such as that above the implication is that secondary sludge re-settles in the same proportion as primary sludge. The estimate of total sludge for treatment is somewhat less with this approach as only a fraction of the secondary sludge is assumed to have re-settled. There is little difference in the estimate of BOD proceeding to the next stage.

10.5 Final settling tanks and thickeners

Where suspended solids are in high concentration the mode of settlement differs from that where the particles are sufficiently far apart for mutual interferences to be negligible. In the mining and chemical industries the need to increase the concentration of a suspension, decanting off the clear liquor, frequently arises and is often accomplished by the use of continuous flow thickeners which resemble circular sedimentation tanks but have to be designed for highly concentrated suspensions. The principles have been familiar to chemical engineers for many years[9] and in cases where a rational approach to design is warranted they are applicable to final (activated-sludge) settling tanks and to sludge thickeners. Since the behaviour of sewage sludges is more variable and less well understood than that of inert slurries, sewage works designers use surface loading and/or detention period values which are known from long experience to be suitable. Nevertheless it is worthwhile for the designer to have some knowledge of the theory of thickeners which is outlined below as rational design is likely to become more common.

The analysis which follows applies to suspensions which clarify in the manner illustrated in Fig. 10.10 which shows successive stages in a settling column test (it should be noted that this behaviour is different from that shown in Figs 10.7 and 10.8). The contents of the column should be gently stirred to avoid wall effects and at the beginning of the test the column is occupied by a uniform concentration of the suspension, Fig. 10.10(a). After a short time, Fig. 10.10(b), a zone of clarified liquor, A, can be observed below which there is zone, B, in which the suspension is at its initial concentration. In many cases, especially when the particles are in a narrow size range, the interface between zones A and B is clearly visible. By this time a zone D of thickened material will have begun to accumulate at the bottom of

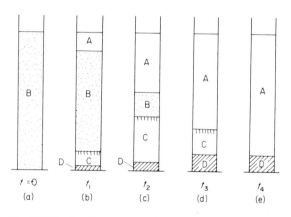

Fig. 10.10 Settling test on a flocculent suspension

the column. In zone C, between B and D, concentration of particles increases from the initial value in B to that of the thickened material in D. Between B and C there is a transition zone. The boundary between C and D is not always clearly defined and often vertical channels can be observed in the transition zone and in the upper levels of C through which water is being forced upwards from the suspension below as it consolidates under gravity.

As time proceeds D grows larger and B, which remains at the initial concentration, becomes smaller, Fig. 10.10(c), and then disappears, Fig. 10.10(d). Eventually only zones A and D remain, Fig. 10.10(e), all the original suspended particles now being in zone D with clear liquor above. Zone D may shrink to some extent as further consolidation occurs. Fig. 10.11 shows the results graphically.

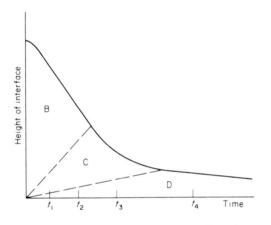

Fig. 10.11 Results of settling test on a flocculent suspension

In this process settling velocity decreases as the local particle concentration increases. Thus between zones B and D, settling velocity decreases in the downward direction.

Figure 10.12 shows zones in a continuous flow thickener corresponding to those in column tests. The suspension enters at the centre of the tank through a baffled inlet and zone B contains suspension at the inlet concentration. Above B is a zone of clarified liquor, A, which forms that part of the outflow which passes over the weir. The remainder of the outflow is withdrawn from the base of the tank as underflow, often at a rate of the same order as the overflow. Initially the tank will fill with a suspension of inlet concentration, but after a period of constant inflow and outflow rate, a steady-state equilibrium condition will be achieved with the several zones as illustrated in Fig. 10.12, provided the tank is of appropriate size in relation to the input flow, the concentration and the underflow.

A mass balance for the suspended material in input and output may be written:

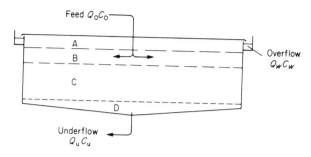

Fig. 10.12 Steady-state zones in a clarifier (or thickener) treating a flocculent suspension

$$Q_0 c_0 = Q_u c_u + Q_w c_w$$

in which Q_0, Q_u and Q_w are the volumetric rates of inflow, underflow and overflow, and c_0, c_u and c_w are the corresponding suspended solids concentrations. Usually c_w is negligible compared with c_0 and c_u, and we can write simply:

$$Q_0 c_0 = Q_u c_u \qquad (10.1)$$

The rate of flow of particles through unit plan area of the tank is known as the mass flux. Thus, the input mass flux is expressed:

$$M_0 = Q_0 c_0 / A$$

and underflow mass flux as:

$$M_u = Q_u c_u / A$$

Mass flux may be expressed as the product of concentration and downward velocity of the particles so at some level, i, in the zone C where concentration increases downwards, mass flux may be expressed:

$$M_i = c_i (v_i + Q_u / A)$$

The bracketted term gives the total downward velocity made up of the velocity v_i with which the particles are settling through the fluid, and Q_u / A, the downward velocity of the fluid in which the particles are suspended.

In hindered settling, the velocity is related inversely to concentration, so proceeding downwards in zone C , as concentration c_i increases, settling velocity, v_i decreases. In consequence M_i may have a limiting value, M_L which will be in the highest rate at which particles can descend to a lower level. The underflow mass flux is then limited to M_L. If the input mass flux, M_0 exceeds M_L, the layer at which M_L occurs will expand upwards and solids of mass flow $A(M_0 - M_L)$ will begin to pass over the effluent weir. The design objective is therefore to ensure that sufficient area of tank is available for M_0 not to exceed M_L.

In order to find whether a limiting value will occur, a relationship is needed between v and c. Settling column tests can be carried out with various dilutions of the suspension in order to obtain velocities of descent of the A–B interface, Fig. 10.11, corresponding to different concentrations. Often, pairs of values of v and c give a straight line when log v is plotted against c, indicating a relationship of the form:

$$v = v_0 e^{-kc}$$

in which v_0 is the intercept when c is zero and k is the slope of the line. Less often both scales have to be logarithmic to achieve a straight line and the relationship becomes:

$$v = Kc^\alpha$$

To illustrate the occurence of a limiting value of mass flux, v_0 will be taken as 8 m/hr and k as 0.4 m³/kg so that

$$M_i = 8c_i e^{-0.4c_i} + c_i Q_u/A \qquad (10.2)$$

The values used for the constants are not untypical but should not be taken as standard.

Fig. 10.13(a) shows v against c. In Fig. 10.13(b), the ordinates of the first graph have been multiplied by values of c so that the graph illustrates the first term of equation (10.2). On Fig. 10.13(c), the second term of the equation is shown by broken lines, the lower one for $Q_u/A = 0.6$ m/hr and the upper one for 1.2 m/hr.

Adding the ordinates of graph (b) of Fig. 10.13 to each of the straight lines gives the whole of the relationship of equation (10.2) for each of the Q_u/A values used.

The lower curve has a minimum of 7.4 kg/(m²hr) at a concentration of 9 kg/m³. This indicates that no more than 7.4 kg/m² per hour can pass through the zone C if the concentration within that zone reaches 9 kg/m³. Consequently if all the solids are to reach the underflow,

$$M_0 = \frac{Q_0 c_0}{A} \nless 7.4 \text{ kg/(m}^2\text{hr)}$$

If the aeration tanks were operating with a mixed liquor solids concentration (MLSS) of 4 kg/m³ (4000 mg/l), the area of settling would have to be such that Q_0/A did not exceed 7.4/4 = 1.85 m/hr.

In determining the area, the maximum rate of flow should be used for Q_0. During sustained high flow the MLSS in the aeration tank will decrease a little, there being a higher proportion of the sludge stock in the final settling tank and a higher sludge blanket, but when inflow to the aeration tanks rises quickly, outflow will also rise while the suspension displaced to the settling tanks will be at its original concentration.

The upper curve of Fig. 10.13(c) has no minimum point and gives a mass flux value of about 11 kg/(m²hr) for a concentration of 4 kg/m³, requiring $Q_0/A = 11/4 = 2.75$ m/hr. However, such a high rate might result in upward velocities in excess of particle settling velocity in the upper part of the

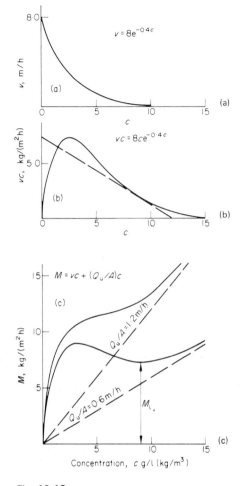

Fig. 10.13

tank and particles being carried over the weir in consequence. The traditional rule for final settling tanks was to allow about twice as much maximum
flow per unit area as for primary tanks which would give a limit of about
2.4 m/hr.

Settling tanks need sufficient area both to allow particles to settle to the
floor and to avoid having upward water velocities towards the surface in
excess of particle settling velocities. In any particular application, one of
these criteria over-rides the other. Primary settlement has often been
regarded as an instance of the second of these criteria while mass flux theory
is concerned with the first. However, as can be seen from the example given,

both criteria may have to be considered if a high underflow rate is contemplated.

It can be shown that the minimum point on the mass flux curve disappears when $Q_u/A = v_0 e^{-2}$ which is 1.08 m/hr when v_0 is 8 m/hr. The higher rates seem to be irrelevant in final tank design because of the risk of excessive upward velocities.

Limiting mass flux rates can be found without constructing curves like those in Fig. 10.13(c) by drawing downward sloping lines tangential to the curve of Fig. 10.13(b). The one for $Q_u/A = 0.6$ m/hr is shown. The limiting value, 7.4 kg/(m²hr), is given by the intersection of the sloping tangent with the vertical axis of the graph.

Using the mass flux theory for design, a trial value of Q_u/A can be used to find the corresponding value of M_L from which the maximum allowable value of c_0 can be obtained from:

$$c_0 = M_L \frac{A}{Q_0} = \frac{M_L}{(Q_u/A)} \cdot \left(\frac{Q_u}{Q_0} \right)$$

in which Q_u/A is the trial value, Q_u is the rate of re-cycle flow and Q_0 is the maximum rate of inflow consisting of the maximum rate of throughput of the works, normally '3DWF', together with the re-cycle flow. For a typical re-cycle flow of 1 DWF and

'3DWF' $= 3(D + E) + I =$ say, 2.6 DWF

for Q_u/A of 0.6 m/hr and a corresponding M_L of 7.4 kg/(m²hr), the maximum concentration would be:

$$c_0 = \frac{7.4}{0.6} \times \frac{1}{(2.6 + 1)} = 3.4 \text{ kg/m}^3$$

Repetition with different values of Q_u/A will establish the relationship with c_0 so that the Q_u/A corresponding to the MLSS of the aeration tanks can be identified and A calculated using the re-cycle flow for Q_u.

An alternative method, avoiding trial, is as follows.

The differential of vc with respect to c can be equated to $- Q_u/A$ to give an equation for c_L, the concentration at which the limiting value of M occurs:

$$Q_u/A = (kc_L - 1)v_0 e^{-kc_L} \qquad (10.3)$$

Substituting this in equation (10.2) gives

$$M_L = v_0 k c_L^2 e^{-kc_L}$$

Dividing the expression for M_L by the one for Q_u/A gives

$$c_u' = kc_L^2/(kc_L - 1)$$

in which c_u' is the underflow concentration when the limit applies, equal to

$$M_L/(Q_u/A)$$

The solution to this quadratic equation in c_L is

$$c_L = \tfrac{1}{2}c_u' (1 + \sqrt{(1 - 4/kc_u')}) \tag{10.4}$$

Neglecting losses over the weir, c_u' may be replaced by Rc_0, in which R is the ratio of the maximum value of Q_0 to Q_u, and the required area can then be found by substituting c_L in equation (10.3).

Taking v_0 and k as above with c_0 as 3.5 kg/m³, Q_u as 500 m³/hr and Q_0/Q_u as 3.6, the following results are obtained: $c_L = 9.162$ kg/m³, $Q_u/A = 0.546$ m/hr and $A = 916$ m².

Both of the procedures illustrated require safe design values for v_0 and k of which there is little experience at present.

The settling properties of activated sludge are measured as a matter of routine because of danger of the development of filamentous organisms which cause a deterioration in settling known as 'bulking', which can be controlled to some extent by changes in the operation of the aeration tanks.

The Water Research Centre has developed a standard apparatus[12] in place of the one litre measuring cylinders used formerly. This consists of a settling column 0.5 m deep and 0.1 m in diameter with a simple 1 rev/min stirrer to eliminate wall effects. The height of the interface below the clarified liquor is measured after allowing 30 minutes for settlement and this is used to find the Stirred Specific Volume Index:

$$\text{SSVI} = \frac{100}{\text{initial conc. as \%}} \times \frac{\text{height of interface after 30 min}}{\text{initial height}}$$

The test is carried out at two or more initial concentrations, obtained by dilution, and the value interpolated for a standard concentration of 3.5 kg/m³. Typical values range from 60 for works producing a well nitrified effluent from domestic sewage to 180 for works producing good though not fully nitrified effluent from sewage containing industrial waste. However, $\text{SSVI}_{3.5}$ may be as high as 200 where an only partially treated effluent is being produced. For use in optimisation studies an empirical relationship has been deduced[8] to enable M_L to be calculated from $\text{SSVI}_{3.5}$.

Commonly used standards for final settling tanks have been to use somewhat lower maximum flow detention periods than for primary tanks and correspondingly higher surface loading rates with the same depth of about 3 m (p. 218). Since the re-cycling of activated sludge makes the maximum flow into final tanks 25 to 30% higher than that for the primary tanks, the total volume is about the same for each set of tanks and similar dimensions have sometimes been adopted.

Humus settlement is less amenable to rational analysis than activated sludge settlement, partly because the release of humus from filters varies seasonally and in other ways. An earlier custom was to use a settling velocity or surface loading rate about twice the 1.2 m/hr used for primary tanks, halving the detention period. More recently design seems to have been more cautious, 2 m/hr being sometimes quoted as a maximum or even the same standard as for primary tanks.

The flow per unit length of weir is considered to be a factor in tank perfor-mance, especially for humus and activated sludge settling. If the flow per unit length is high there may be a risk of solids being drawn over the weir by the concentration of flow. Opinions as to safe maxima vary from 5 to 20 m³/hr per metre length of weir. Large diameter tanks have less circumference in relation to flow than small tanks since circumference increases only directly with diameter while flow — assuming depth remains the same — increases with the square of diameter. For this reason large tanks have weir troughs a short distance from the outer wall to provide extra weir length. Vertical pipes at intervals convey effluent from the trough. The annular space between the trough and the outer wall is swept independently and has its own sludge pockets or Vee scrapers are employed.

10.6 Division of flow between tanks

Where a continuous flow treatment process occurs in several identical tanks operated in parallel, it is immaterial whether the tanks take equal shares of the flow provided that the process is linear with time, over the range of deten-tion times which occur. If the process is non-linear with time, the best result in terms of the quality of the effluent from the plant can be achieved only if the load is shared equally among the units. Since sewage sedimentation is non-linear with time, it is usually considered to be worthwhile to design feed channels and pipes so that flow is divided equally between the tanks of a group.

With a group of circular tanks, this may be done as shown in Fig. 10.14. The channels bifurcate successively and the lengths and levels of corres-ponding branch channels are equal. The channels terminate in feed wells each of which serves four tanks. Penstocks are provided in the feed wells so

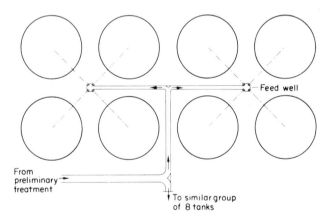

Fig. 10.14 Symmetrical arrangement of circular tanks to equalise flow

that individual tanks may be taken out of service for maintenance. If the flow to a single tank is shut off, the remaining tanks will not take perfectly equal shares of the total input but, provided that the major proportion of the total head loss occurs in the feed pipes rather than in the channels, the inequality will be trivial. This is one reason for using channels, a second reason being that, since the depth of channel flow varies with discharge (it is controlled primarily by the feed pipes), velocity variations may be narrower than discharge variations.

The principle of Fig. 10.14 can of course be used for upward-flow and rectangular tanks.

Where site restrictions would make a symmetrical layout very costly, the principle of successive bifurcation may have to be abandoned. If the tanks are fed by branches from a common channel, it is possible to proportion channels and pipework in such a way that flow is equally divided between tanks. Strictly, this can be done for one rate of flow only, but if the losses in the channels are small compared with those in the pipes, the feeds can be nearly equalised over a reasonably wide range of flow. If the channel cross-section can be made large enough there may be no need to design the feed channels as a manifold system. If necessary, fairly large channels with low velocities of flow can be used so that the channels are virtually long reservoirs with little difference in surface level. Material is kept in suspension by air bubbles from diffuser domes (see Section 11.8) installed on the floor.

Unless diffused-air plant or other means (e.g., surface aerators) of keeping material in suspension is to be installed, channels feeding sedimentation tanks should be designed so that the mean velocity does not fall below 0.3 m/s except for short periods. Lower velocities at minimum flow may have to be accepted for feed pipes in order to limit friction losses at maximum flow. The velocities at high flows will be relied upon to scour deposits. The hydraulic design of works is discussed further in Section 14.4.

References

1 Clements, M.S., Velocity variations in rectangular sedimentation tanks, *Proc. Inst. Civ. Engrs*, **34**, 171–200, June 1966.
2 Hubbell, G.F., Hydraulic characteristics of various circular settling tanks, *J. Am. Wat. Wks Assn*, **30**, 335–353, 1938.
3 Townend, C.B. and Wilkinson, G.W., Some hydraulic aspects of sewerage and sewage disposal, *Proc. Inst. Civ. Engrs*, Pt III, **4**, 662, Dec. 1955.
4 Walton, R. and Key, T.D., Application of experimental methods to the design of clarifiers for waterworks, *J. Inst. Civ. Engrs*, **13**, 21–48.
5 Vokes, F.C. and Jenkins, S.H., Experiments with a circular sedimentation tank, *J. Inst. Sew. Purif.*, 11–24, 1943.
6 Steele, B.D., The treatment of storm sewage, *Symposium on Storm Sewage Overflow*, Inst. Civ. Engrs, London, 1967.
7 White, J.B. and Allos, M.R., Experiments on wastewater sedimentation, *J. Wat. Pollut. Control Fedn*, **48** (7), 1741–1752, July 1976.

8 Water Research Centre, *Sewage Treatment optimising model — user manual and description* TR 144, WRC, Stevenage, March, 1981.

9 Coe, H.S. and Clevenger, G.H., Methods for determining the capacities of slime settling tanks, *Trans. Am. Inst. Min. Engrs*, **55**, 356–385, 1916.

10 Kynch, G.J., A theory of sedimentation, *Trans. Faraday Society*, **48**, 166–187, 1952.

11 Dick, R.I., Evaluation of sludge thickening theories, *Proc. Am. Soc. Civ. Engrs*, **93**, 9–92, August 1967.

12 White, M.J.D., Settling of activated Sludge, *Tech. Report TR 11*, Water Research Centre, 1975.

11

Biological oxidation processes

11.1 Introduction

An outline of the biological oxidation processes and of their relation to the cycle of growth and decay has been given in Chapter 8. The commonly adopted processes will now be taken in turn so that the fundamentals can be discussed in more detail and accounts given of the design of plants.

Activated-sludge treatment

11.2 General description

The components of an activated-sludge plant are the aeration tanks in which the biological oxidation takes place, the settling tanks for the recovery of the activated sludge, and a system of pipes and pumps to return the activated sludge to the inlet end of the aeration tank.

The input to the system consists of sewage or other wastewater which has usually received treatment in primary sedimentation tanks. In a municipal treatment plant, the waste entering the aeration tanks has a BOD of 200 to 300 mg/l and a suspended solids concentration of 100 to 150 mg/l. The flow of returned activated sludge entering the aeration tanks in a modern plant is of the same order as the mean flow of settled sewage (i.e. 1 × DWF) though in earlier plants it was only 20% to 30% of the mean waste flow. The mixture of settled sewage and returned activated sludge has a suspended solids concentration in the range 2500 to 4000 mg/l.

The aeration tank may be designed to operate on the 'plug flow' principle in which the mixed liquor steadily moves from one end to the other, or it may be a form of mixed system in which influent is dispersed to all parts of the tank (see Section 11.5).

Air diffusers or mechanical surface aerators installed in the aeration tank maintain a dissolved oxygen concentration up to 1 or 2 mg/l and serve also to keep the activated sludge in suspension and well mixed with the waste. The aerobic organisms in the activated sludge use the organic material in the waste as a source of food, partly to maintain life and partly to synthesise new cell material. Typically 0.5 to 1.0 kg of activated-sludge suspended solids is produced for each kg of BOD removed.

The treated mixed liquor passes to the final settling tanks. In normal circumstances, activated sludge settles quite readily leaving a clear super-natant liquor which is the end product of the conventional sewage treatment process.

Typically the suspended solids concentration of the effluent is 20 to 30 mg/l, the solids being mainly activated sludge and therefore containing living organisms. The BOD is usually between 10 and 20 mg/l. Part of this is exerted by the suspended solids, part by organic matter not metabolised during treatment.

11.3 Biological aspects of activated sludge

The living organisms in activated sludge form a complex ecosystem depen-dent on the organic material in the sewage for their nutrients and energy, Fig. 11.1. The main primary feeders are bacteria, present in colonies which form the floc particles. Heterotrophic bacteria obtain their nutrition from complex organic compounds in solution using them partly to produce energy by a respiratory process which requires oxygen and converts organic material into carbon dioxide and water. Some of the energy is used to synthesise new cells from the nutrient material. Autotrophic bacteria use carbon dioxide as their source of carbon and can oxidise inorganic compounds. Their presence enables nitrites and nitrates to be produced by the oxidation of ammonia.

In addition to the bacteria there is a varied fauna of protozoa. Of the order of 100 different species have been identified in activated sludge. They range from rhizopods (like amoeba) through flagellates (like euglena) and ciliates (like paramecium) to stalked species (like the vorticella). Some of the species are primary feeders but for the most part the protozoa prey upon the bacteria performing the useful role of limiting the bacterial population to keep it in

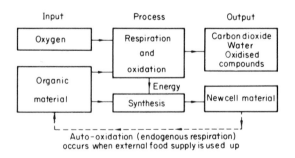

Fig. 11.1 Action of aerobic organisms

balance with the food supply. The protozoa are mainly attached to the floc but some are free-swimming.

Higher still in the ecosystem are rotifiers and nematode worms. Larger and much less numerous they prey upon the protozoa.

All these organisms are growing and multiplying while food and oxygen are available but the total stock of sludge is kept constant by removing surplus microbial mass. Consequently the mean residence time of the sludge in the plant may be estimated by dividing the total mass of activated sludge by the rate of growth or by the total rate of wastage (which includes that lost with the effluent). This time period is known as the 'sludge age' or 'mean cell residence time'.

Sludges of low age have smaller floc particles, and may be difficult to settle. They contain few of the higher forms of life which only have chance to develop in older sludges. The predominant species of protozoa vary with sludge age in the sequence in which they were mentioned above.

The autotrophic bacteria responsible for nitrification grow more slowly than the bacteria which perform most of the carbonaceous oxidation. Consequently little nitrogenous oxidation can be expected from a sludge of low age.

Fig. 11.2 shows how the mass of micro-organisms increases when they are placed in a nutrient solution. In the early stages when food is abundant, growth is exponential. This is followed by a constant growth phase and then, as food becomes more scarce, by a phase in which the rate of growth declines. Eventually when some essential nutrient is exhausted growth ceases and life can continue only by the microbial mass using its own internal reserves. The mass declines during this 'endogenous' phase.

A continuous process with wastage of microbial mass to compensate for growth could in principle be operated on a small section of any part of this curve. Along the steep part of the curve the rate of removal of organic

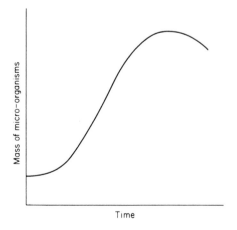

Fig. 11.2 Growth of mass of micro-organisms

material from solution would be very high as also would be the growth. However, it would be necessary to maintain a high concentration of nutrients and then the effluent would contain a high concentration of unmetabolised organics. Also as sludge age would be low difficulty might be experienced in recovering it by settlement for re-use.

Conventional activated-sludge plants work instead in the declining growth phase. It is possible to operate over into the endogenous phase so that net growth is zero and there is no surplus sludge, other than accumulated non-biodegradable matter, to be disposed of. However, increased requirements for energy, and volumetric capacity of plant render 'extended aeration', as this form of treatment is called, uneconomic except in special circumstances.

Occasionally 'bulking' of activated sludge occurs. Bacteria are then present in the form of filamentous growths and the sludge becomes difficult to settle in the final clarifiers. The causes of bulking are still not fully elucidated but it has sometimes been observed that the condition is associated with low levels of dissolved oxygen in the aeration tanks. Another possible cause is some form of nutrient deficiency in the waste being treated.

11.4 Range of application

Most commonly in municipal plants the activated-sludge process is used to remove BOD from previously settled sewage to produce a final effluent of about 20 mg/l. However, it can be applied in a number of ways.

If a fully nitrified effluent is required rather more BOD removal must be ensured and further aeration must be provided, for oxidation of the nitrogenous material. Since this requires larger tank capacity and more power an alternative is to use a two-stage process in which the second stage is mainly for nitrification. Though it is possible in principle to use activated sludge for the second stage (with its own sludge stock) biological filters are more usual. If a second stage is to be used the first stage may be reduced in size so that it removes a smaller fraction of the BOD. The first stage is then termed 'partial treatment' which might conceivably be used alone if the receiving water could tolerate an effluent of relatively high BOD.

All these versions of the process are used with the biomass in the growth stage so surplus sludge is produced and requires disposal. Using the endogenous stage (extended aeration) the net growth of biomass is practically nil, but some sludge still has to be removed from the stock since the inert portion of the suspended solids entering the aeration tank would otherwise accumulate.

Any activated-sludge process can treat screened and degritted sewage without primary settlement but since this might cause even more surplus sludge to be produced when operating in the growth phase it is usually confined to extended aeration in oxidation ditches and small 'packaged' plants where reducing sludge disposal to a minimum compensates for the extra aeration capacity and power needed.

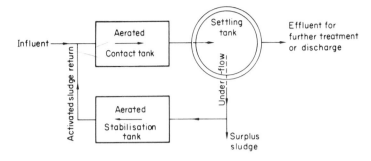

Fig. 11.3 Contact stabilisation

A quite different variant is contact stabilisation, Fig. 11.3. The waste to be treated is mixed with returned activated sludge in a contact tank with a detention period of 20 to 30 minutes. Little oxidation is achieved at this stage but the particulate matter in the waste is flocculated by the sludge which is then separated in a settling tank. The returned sludge flow passes through a stabilisation tank in which the BOD of the particulate matter carried by the sludge is satisfied. The clarified outflow from the settling tank contains most of the dissolved matter originally in the waste so that contact-stabilisation is mainly of use when only a small part of the BOD is accounted for by organics in solution. However, if necessary, further treatment can be applied to the settling tank effluent and indeed 'bio-flocculation', as the contact-stabilisation process was originally called, was used many years ago as a pre-treatment process to relieve the load on biological filters.

A more recent development is the provision of anoxic zones which are sections of tanks without aerators but with sufficient turbulence to maintain sludge in suspension. Both settled sewage and activated sludge enter the tank in the anoxic zone. Because DO levels are low, the heterotrophic bacteria in the activated sludge are forced to obtain their oxygen from the nitrate in the treated liquor carrying the sludge, free nitrogen being released to the atmosphere. Thus oxygen originally used for nitrification is re-used in BOD removal, nitrate being removed from the effluent in the same process. Some energy is saved as the net amount of oxygenation is reduced. The possibility of de-nitrification in final settling tanks is reduced so the activated sludge is less likely to become buoyant through the evolution of nitrogen bubbles.

The range of operating conditions in trade-waste treatment plants is much wider than in municipal sewage treatment plants, higher BOD being common for trade-waste treatment. Wheréver organic matter can be bio-oxidised an activated sludge (or a biological filtration) process can be used. This applies to synthetic organics as well as waste vegetable matter and the like. The waste must be free from toxic material and from any inhibitory matter to which the sludge cannot be acclimatised. If the waste is deficient in nitrogen or phosphorus, salts containing these elements must be added to give C : N : P ratios of about 150 : 5 :1. Domestic sewage is usually needed for starting up and may be needed continuously as a small proportion of the total flow.

11.5 Plug flow and mixed systems

In a plug-flow system the waste to be treated is joined by recycled activated sludge at the inlet end of the tank and the mixed liquor moves steadily to the outlet end without any longitudinal intermixing of the tank contents. Baffle walls across the tank have often been used to prevent short-circuiting and longitudinal intermixing.

In a completely mixed system some means is adopted for ensuring that the whole of the tank contents are continuously and thoroughly mixed. With such a system each 'parcel' of input is rapidly and uniformly dispersed throughout the tank. The contents of the tank are thus of uniform composition throughout consisting of a mixture of elements at various stages of reaction. The effluent has a composition identical with that of the contents.

A small, single unit tank is inevitably a mixed system. A large tank would need multiple inlets and outlets together with efficient stirring to be completely mixed. However, it is more common to achieve mixing by using a tank in the form of an endless loop around which mixed liquor circulates at a rate considerably higher than the throughput. Input enters at only one point but makes on the average many circuits during its residence time, intermixing throughout with the tank contents. With a high degree of circulation operation approximates to that of an ideal dispersion tank.

With plug flow the whole of the input does not necessarily have to be introduced at the head of the tank and indeed most plants have provision for dividing the input among several inlets at points along the first third or half of the tank. The objective of this incremental-feed or stepped-aeration system is to enable the activated sludge to take up load progressively. Provision of several inlets also enables the first part of the tank to be used for re-aeration of recycled sludge before it is contacted with the waste.

Since with plug flow assimilation occurs steadily as the mixed liquor moves along and since the reaction rate is dependent on concentration remaining, the need for oxygen is high at the inlet end and progressively less towards the outlet. Consequently the degree of aeration should vary along the tank to save energy. Provision for this is known as tapered aeration.

Since substrate concentration is higher in most of a plug-flow tank than in a mixed tank it might be expected that the mean reaction rate would be higher and that a smaller volume would be needed but there is no convincing experimental evidence that this is the case.

The case in favour of the mixed system rests upon its ability to cope with shock loads since they are quickly dispersed, and upon its generally more stable operation under the diurnal variations of load. It is simpler to control but it is sometimes asserted that mixed systems are more prone to sludge bulking than plug flow systems.

It should be noted that some degree of mixing inevitably occurs even in a plug-flow plant since a significant flow of effluent is returned carrying the activated sludge. In modern plants the recycle flow is of the same order as the net throughput (DWF). Thus, although there is no circulation within the tank there is significant circulation by way of the external returned sludge circuit.

Though the designer can readily provide for incremental feed and tapered aeration in a plug flow system the operator faces many practical difficulties in discovering an optimum set of conditions, and the possibility of making best use of the operational flexibility provided is probably not high.

11.6 Activated-sludge process parameters

Activated-sludge treatment involves maintaining a stock of micro-organisms which is large enough to assimilate the daily quantity of organic material carried by the wastewater. The various forms of the process can be characterised numerically by the relationship between the rate of supply of organic material and the mass of micro-organisms which it is feeding. Neither of these factors can be readily measured but the daily BOD load can be used as an indication of the food supply and the total mass of mixed liquor suspended solids can be used as a measure of the biomass. The ratio of these is known as the food/micro-organism ratio (F/M) or as the 'sludge loading index' and may be expressed as:

$$\frac{Qb_r}{Vs_r}$$

in which Q is the mean rate of flow of wastewater, b_r is the BOD concentration removed from it, V is the volume of the aeration tank and s_a is the mixed liquor suspended solids concentration (MLSS). With volumes measured in Ml and concentration in mg/l, Qb_r gives kg of BOD removed per day while Vs_a gives kg of activated sludge solids. With b_1 and b_2 as BOD of input and output respectively, $b_r = (b_1 - b_2)$ and since, except for partial treatment, b_2 is very small compared with b_1 the food supply is often expressed as Qb_1 and may be obtained alternatively as the product of the population served and the daily BOD per head (about 0.055 kg per head per day). This product is reduced by 30 to 40% if the sewage receives primary settlement (see p. 219). Trade wastes may be taken into account by regarding them as equivalent to additional population.

It should be noted that strictly b_2 is the unassimilated BOD. Effluent BOD will be greater than this due to the presence of the small quantities of suspended matter which the final settling tank has failed to remove.

In municipal sewage treatment F/M ratios of 0.15 to 0.30 are commonly adopted, the lower end of this range being used if a nitrified effluent is aimed at. Extended aeration is associated with lower F/M down to about 0.05. With a ratio of 0.5 or more the process is termed high rate and at higher values significant BOD concentration remains in the effluent so that the process is appropriate only where a partially treated effluent is required, for example, where high-rate nitrifying filters are to be used as a further process.

The 0.3 to 0.4 range is sometimes avoided as poorly settling sludges have been experienced within it. These are also found, though possibly for different reasons, in the range between extended aeration and 'conventional' treatment. However, overall economic factors often point distinctly either to extended aeration or to conventional treatment as explained later.

Table 11.1 is given to illustrate the contrasts between various forms of treatment and to show how the several factors can be estimated. Values are given for F/M, line (1), appropriate to extended aeration (0.05), medium rate (0.15) and high rate (0.5) treatment. The rest of the table is worked out for a mean flow, Q of 10 Ml/d, b_r of 200 mg/l and s_a of 4000 mg/l, though it must be emphasised that the same removal of BOD would not necessarily be achieved by each of these processes with a given waste.

Line (2) shows the considerably greater volume needed for extended aera-

Table 11.1

(1) F/M kg BOD per d/kg MLSS	$\dfrac{Qb_r}{Vs_a}$	0.05	0.15	0.50
(2) Vol. of tank, m³	$\dfrac{Qb_r}{(F/M)s_a} \times 10^3$	10 000	3333	1000
(3) Detention time, h	$\dfrac{b_r}{(F/M)s_a} \times 24$	24	8	2.4
(4) Sludge yield coeff., Y kg surplus/kg BOD	values assumed for illustration	0.4	0.8	1.0
(5) Surplus sludge kg/d	YQb_r	800	1600	2000
(6) Oxygen coeff., m kg O₂/kg BOD rem.	values assumed for illustration	2.5	1.2	0.7
(7) Mass of oxygen/day kg/d	$mQ\,b_r$ $(+\,4.5\,Qc_n)$	5000	2400 (+900)*	1400
(8) Volumetric loading kg BOD per day/m³ of tank	$\dfrac{Qb_r}{V} = \dfrac{F}{M}s_a \times 10^{-3}$	0.2	0.6	2.0
(9) Aeration per unit vol. kg O₂ per day/m³	$\dfrac{mQb_r}{V} = \dfrac{F}{M}s_a \times 10^{-3}$	0.5	0.72 (+0.27)*	1.4

*for 20 mg/l ammoniacal nitrogen (c_n) to be oxidised

tion, implying a correspondingly larger stock of sludge since in this illustration the same value of s_a is used throughout. The cost of the tanks will not usually increase in proportion to the volume since the form of construction of extended aeration plants (such as oxidation ditches) is simpler than that of medium and high rate plants.

Detention time, line (3) can be expressed as V/Q. It is the same as the mean contact time which can be shown to be independent of and therefore uninfluenced by the sludge recirculation rate.

The sludge yield coefficients in line (4) have been assumed for purposes of illustration. They are not untypical and reflect the high degree of endogenous respiration which can be expected at low F/M where cells synthesised from oxidation of the organics in the waste are themselves oxidised. However, the values given should not be taken to be general. Laboratory or pilot-scale work would be needed to obtain reliable estimates for a given case.

Line (5) indicates the main benefit of extended aeration in that surplus sludge production is minimised thus reducing disposal costs. The figures are for total sludge production. If, say, 20 mg/l of suspended solids remain in the final effluent the surplus sludge requiring disposal will be 10 (Ml/d) × 20 (mg/l) = 200 kg/d less than the figures given in the table.

As in line (4) fairly typical values have been assumed for line (6). Again, experimental work would be needed in a given case. Note that the large sludge stock at low F/M needs more oxygen relative to the BOD removal.

Line (7) follows from line (6) and may be used to estimate the energy needed to provide aeration. About 0.6 kW h would be needed for each kg of oxygen supplied through aeration. If nitrification were to occur (final BOD of about 11 mg/l would have to be achieved) further oxygen would be needed to the extent of about 4.5 times the mass of ammoniacal nitrogen to be oxidised. Thus, for 20 mg/l of ammoniacal nitrogen in the 10 Ml/d flow an additional 900 kg/d of oxygen would be required.

Volumetric loading, line (8), is often used instead of F/M as a parameter to characterise the form of treatment, though it needs to be accompanied by the value s_a.

Oxygen required per unit volume, line (9) is useful as an indication of the intensity of aeration required from the equipment to be installed. It should be noted that though low F/M requires more oxygen, the volumetric intensity is less implying that aerators can be more widely spaced. Additional aeration for nitrification is again given in the 0.15 column.

The oxygen requirements in lines (7) and (9) are daily averages. Installed capacity would need to be high enough to serve the daily peak load.

The figures in lines (2), (3), (8) and (9) depend on s_a. With a value different from the 4000 mg/l assumed, lines (2) and (3) would change in inverse proportion to s_a while those in lines (8) and (9) would change in direct proportion. Values of s_a typically range between 2000 and 4000 mg/l but are sometimes as high as 6000 or even more. A high value, if feasible in the particular circumstances, will reduce the size of tank but will demand a higher volumetric intensity of aeration, larger final settling tanks (if designed rationally) and possibly a higher recirculation rate. Though s_a has to be assumed in design, it is under the control of the operator once the plant is in operation. Ideally the operator will find a level which minimises running costs without risking a poorly settling sludge.

A further useful parameter is sludge age, t_s. Fundamentally this is the residence time of activated sludge in the plant and is found by dividing the mass of sludge by the daily wastage (which is equal to the daily surplus if the stock is kept constant). In the above notation $t_s = Vs_a/YQb_r$. Less direct measures of sludge age are Vs_a/Qs_1 in which s_1 is the input suspended solids concentration and the 'BOD sludge age', Vs_a/Qb_r which is merely the reciprocal of F/M. Some workers use volatile suspended solids for s_a in these expressions as a more direct measure of active biomass.

Choosing which form of the process to adopt in a given case can usually be done without detailed economic comparison. Limitations of site area and the need to keep down energy costs will point to a high rate process for a large city. A lower F/M with provision for nitrification would be chosen if a high standard effluent were needed. Relatively high surplus sludge production can be accepted because on a large plant economies of scale will be realised in disposal costs. For a small flow in a more rural area extended aeration may become a better proposition. Sludge disposal will be minimised and can be dealt with by transporting surplus to a larger plant thus simplifying operation. For a trade effluent, higher power costs may not be serious since the power needs of the effluent plant will often be very small compared with those of the factory.

The choice of process has an influence on decisions as to the extent of primary treatment. With a high rate process BOD not removed in primary settlement is converted to surplus secondary sludge on a roughly one-to-one basis, and secondary sludge is more difficult to de-water than primary sludge, so reducing primary sedimentation capacity can easily increase rather than decrease sludge disposal costs. With extended aeration on the other hand much less surplus sludge is produced by BOD removal so elimination of primary sedimentation often reduces overall sludge production. More aeration capacity will be needed but increasing the volume of aeration tank (or ditch) may be cheaper than providing primary tanks even when additional power needs are taken into account. Screens and grit tanks at the works inlet can also be omitted, grit being allowed to accumulate in the oxidation ditch and screening deferred until after aeration, which results in cleaner screenings.

Once the form of process has been chosen laboratory or pilot-scale studies should ideally be carried out with the waste to check feasibility and to enable design parameters to be estimated. Mathematical models (see Section 11.12) assist in this work. Finally the designer may feel it prudent to provide margins of extra capacity in aeration tanks, aeration equipment, final settling tanks and sludge return facilities to cover variability of load and possible future changes.

It must be emphasised that *F/M* and related parameters are relatively crude measures. The relationship between the true food/micro-organism ratio and the ratio of BOD/d to MLSS will vary from one waste to another and even vary for the same waste for different *F/M*. It is for this reason that more detailed studies are needed to produce a reliable and economic design so far as the complexities of the process allow.

11.7 Variability of load

The sources of variability are as follows:

1 Diurnal variations of flow and strength in dry weather. Each may be of the order of twice the mean during the middle part of the day quadrupling the loading rate (product of flow and concentration). In addition there may be minor weekly and seasonal cycles.
2 Increased flow during wet weather where the sewerage system is combined giving a maximum flow of '3 × DWF' through the oxidation stage. Except during the early part of a storm when accumulated deposits may be scoured from the sewers, storm flow will not increase load because strength will be reduced by dilution. Indeed some pollution load will be lost by spill from overflows.
3 Random fluctuations superimposed upon the cyclic variations together with occasional, sometimes accidental, discharges of trade waste to the sewers.

The oxidation stage is not subject to the full range of these variations since balancing occurs in the primary sedimentation stage through dispersion and intermixing. Furthermore, outflow from the primary tanks responds fairly quickly to changes in inflow but the effluent displaced by a change in inflow consists mainly of liquor which entered the tank some time previously.

The oxidation stage itself has a balancing influence particularly in a mixed system and recirculation of activated sludge has a similar effect even in plug flow. Since the total detention time of sedimentation and aeration tanks is of the same order as the dry weather periods of high flow, as the duration of the early parts of storm flows and as the duration of many random fluctuations, the sludge stock is not subject to very great fluctuations in load.

Even the reduction of detention time to about one-third of its average value during a prolonged storm may have little effect on effluent quality as the strength of the inflow is also reduced to one-third (or less) of the average. This may be confirmed from the mathematical expression given in Section 11.12.

For these reasons the long standing practice of basing tank volume (or nominal detention period) on daily average loading rather than on maximum instantaneous loading appears to be sound. Estimates of daily energy usage and of surplus sludge production can also be based on mean loading. However, it is found that more intensive aeration has to be applied at times of high loading in order to maintain a high enough DO. Installed capacity of aeration machinery is based therefore on maximum load and varies from 2 or 3 times the mean requirement, on a very small plant, to about 1.3 times the mean requirement on a large plant where more smoothing of variations can be expected in the sewerage system.

11.8 Aeration equipment

Nearly all aeration equipment is of proprietary manufacture. Detailed design is done by the makers and new developments both major and minor occur fairly frequently. The form of tank needed differs according to the system used, so at an early stage in the design a provisional choice has to be made, guided by data or proposed designs provided by manufacturers based on experience with their more recent installations. Once a choice has been made detailed design of tanks and decisions as to the overall capacity are made in consultation with the equipment manufacturer. As design details change fairly frequently only the main features of the several alternatives currently available will be described.

Diffused air (Hawker-Siddeley Water Engineering Ltd). The modern equipment has developed from that used when activated-sludge treatment was first invented. Air is supplied in the form of very fine bubbles from fused aluminium oxide ('Alundum') diffusers attached to pipes laid on the tank floor (Fig. 11.4). Blowers in a building near the tanks provide compressed air. The spacing of the pipes and of the diffusers on the pipes is arranged to give the necessary degree of aeration. In plug flow plants a measure of tapered aeration is provided by closer spacing of diffusers near the inlet end of the tanks where oxygen demand is high. Variations in aeration intensity to match variations in load can be made simply by varying the supply of compressed air. Automatic control can be arranged by having DO probes in the tank.

The plant can be laid out in a variety of ways depending on the proportions of the site available. Feed channels are often arranged so that incremental feed can be used optionally and so that part of the capacity can be used for re-aeration. Diffused air can also be applied to a completely mixed principle of operation.

Periodically diffusers have to be taken out of service and treated in a furnace to remove dust particles from the air which progressively blocks their pores. A spare set of diffusers, sufficient for one unit of the plant, is carried in stock so that units can have their diffusers replaced with cleaned ones in rotation. Air filters provided in the blower house intake considerably reduce the need for diffuser cleaning.

Surface aerators. These may be divided into two classes: those rotating on a vertical axis (e.g., Simplex and Simcar) and those rotating on a horizontal axis (e.g., TNO and Mammoth Rotor). In each case the lower part of the rotor is below the surface of the liquor in the tank. Liquor is scooped up by

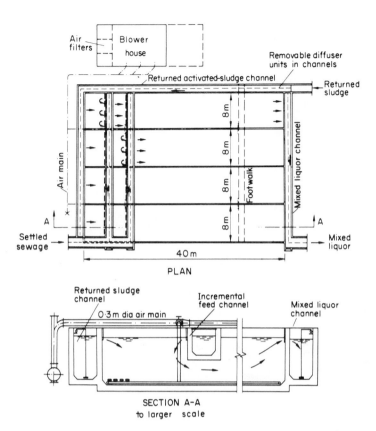

Fig. 11.4 Activated sludge (diffused air) plant (*Hawker Siddeley Water Engineering Ltd*)

the rotor and propelled across the tank. New surface is thus being continuously formed for the diffusion of air, also the tank contents are being circulated at a high rate promoting mixing and keeping sludge in suspension.

The degree of aeration and power absorbed depends on the depth of submergence of the blade tips and can be controlled by raising or lowering either the rotor or the liquor surface level. For tapered aeration in plug flow plants the rotor level is adjusted, rotors at the inlet end being more deeply immersed than those near the outlet. On vertical axis machines this is done by adjusting the lengths of the rods connecting the rotor to the drive. Horizontal axis machines are usually applied to mixed systems where depth of immersion is common to all rotors. The rotors are then installed with their axes across the tank so that they serve to propel the liquor round the circuit.

Varying the degree of aeration to match oxygen demand is achieved by changing the liquor surface level with the aid of some form of adjustable

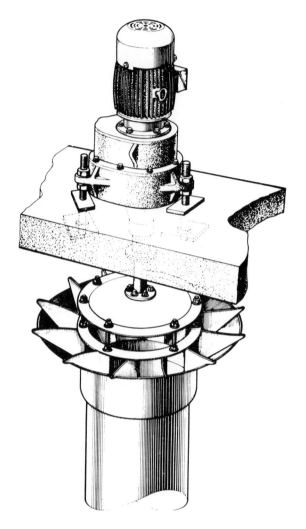

Fig. 11.5 Simplex aerator (*Ames Crosta Babcock Ltd*). In recent designs there is a splash cover above the vanes

outlet weir. In recent installations the weir is automatically raised or lowered in response to the signal from a DO probe in the tank.

Rotors are driven by individual electric motors through reducing gearboxes. The latter account for a large proportion of the capital cost and are the most crucial part of the mechanical equipment as regards maintenance and reliability since they have to perform an arduous and continuous duty.

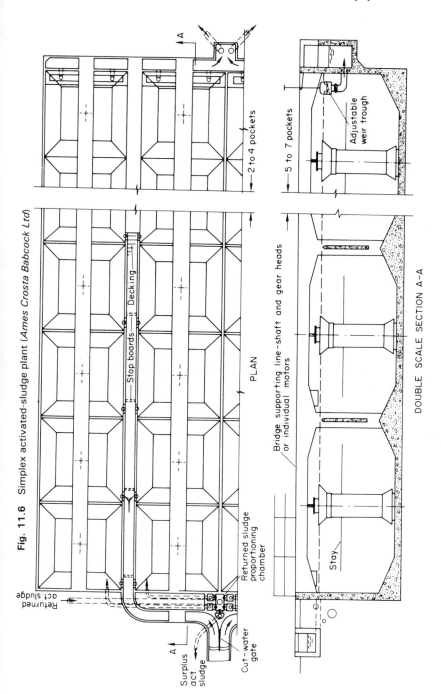

Fig. 11.6 Simplex activated-sludge plant (*Ames Crosta Babcock Ltd*)

(1) Simplex (*Ames Crosta Babcock Ltd*) aeration cones

These have a history of more than 60 years and have been under continuous improvement and development throughout the period. The aeration tank consists of a series of 'pockets', square on plan, each with a rotor at its centre. The typical plug-flow Simplex plant of which there are numerous examples consists of parallel lines of 6 to 10 pockets each. The first few pockets are divided by baffle walls extending to about two-thirds of the depth to prevent short-circuiting and to enable incremental feed or re-aeration to be practised. The feed channel runs alongside the aeration tank with an inlet penstock to each of the first three or four pockets. The floors of the pockets are usually in the form of an inverted truncated pyramid so that dead spaces are avoided but as more powerful aerators have been developed the sloping areas have been reduced and modern examples have entirely flat floors.

The rotor consists of a sheet metal cone carrying shaped blades. The rotor is supported by rods from a driving ring turned by the output shaft of a gearbox mounted on a bridge over the tank. Originally the bridges were longitudinal and carried line shafts which drove several cones from a single motor at the end of the tank. Rotors are now driven by individual motors and it is often more convenient for access to run the bridges laterally across the tanks since longitudinal access can be gained via the feed channels, which are covered.

The trend of development has been towards larger and more powerful aerators so that a modern design would consist of only a few large pockets and the above description would now apply only to a very large plant. A smaller plant can consist of a nest of four pockets arranged for optional mixed or plug flow operation. The total capacity might be divided among two or more such units to provide some flexibility for maintenance.

The rotor draws liquor from an uptake tube to ensure thorough mixing and circulation. In recent plants it has been found feasible to dispense with the tube as shallow tanks are now more usual though tubes would still be used with deeper tanks.

Simplex cones mounted on floats have been used for aerated lagoon treatment and for aerating rivers near to points of effluent discharge.

(2) The Simcar (*Simon-Hartley Ltd*)

Shown in Fig. 11.10 the Simcar is a more recently developed vertical axis rotor. The blades are on the outside of an inverted cone, flow taking place entirely on the outside of the cone surface in contrast to the Simplex which having been conceived with an uptake tube has blades on the inside of a truncated cone so that flow is drawn through the cone from the tube. Since both these aerators have vertical axes much of what has been said about the arrangement and general form of tanks applies to both of them.

(3) The TNO (*Whitehead and Poole Ltd*)

The TNO rotor developed from the Kessener brush rotor introduced more than fifty years ago. The original form of Kessener plant consisted of a rectangular tank with the brushes running along one side propelling flow across the surface so that circulation was in the lateral plane. With plug flow longitudinally the net effect was that flow followed a helical path down the length of the tank. A vertical baffle below the rotor formed a space next to the side wall so that liquor was drawn up from the lower part of the tank as in the Simplex uptake tube.

The brushes were superseded by rotors with comb-like paddles and then, as the result of experiments at the TNO laboratories in The Netherlands, by angle-section blades bevelled at the ends. A parallel development was the application of TNO rotors to the oxidation (or Pasveer) ditch. This is a form of completely mixed system. Liquor circulates continuously in a channel which is oval in plan and trapezoidal in cross-section. The rotors which are supported across the ditch serve to move the liquor around the circuit as well as to provide aeration. The rate of flow around the circuit (at least 0.3 m/s) is much higher than that of the throughput, elements of input making on the average many circuits during their residence time and complete mixing

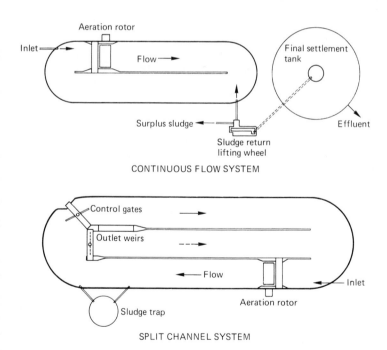

Fig. 11.7 Oxidation ditches using TNO aerators (*Courtesy Biwater Sewage Treatment, Whitehead & Poole Ltd*)

Fig. 11.8   TNO rotors in oxidation ditch (*Courtesy Biwater Sewage Treatment, Whitehead & Poole Ltd, and Central Regional Council*)

being achieved. In principle there is a similarity to the 'Sheffield' system which will not be described since it is unlikely to be used for new work though existing plants continue to be of service.

Oxidation ditches usually have long detention periods (24 hours or more) being in effect extended aeration plants and are mostly used for rural communities and trade wastes. Where sewage is delivered intermittently from a pumping station it is feasible to design the ditch for intermittent operation, the rotors being stopped to allow settlement after which clear liquor is drawn off. This avoids the need for a separate final settling tank.

The oxidation ditch is now the most common application for TNO rotors, Fig. 11.8.

(4) The Mammoth Rotor

The Mammoth Rotor and other types of large, horizontal-spindle aerators are used in tanks (Fig. 11.9) similar in principle to the oxidation ditch but the scale is larger and the detention period and F/M are often in the medium rate rather than extended aeration range though extended aeration can be used with this equipment if it is desired to minimise surplus sludge production. The tank is rectangular on plan with splayed corners and a central baffle wall. Rotors span between the outer wall and the baffle, promoting rapid circulation of the liquor to give a mixed system.

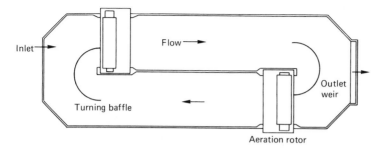

Fig. 11.9 Aeration plant with 1.0 metre diameter horizontal rotors (*Courtesy Biwater Sewage Treatment, Whitehead & Poole Ltd*)

(5) The Carrousel (*Simon-Hartley Ltd*)

The Carrousel is a version of the mixed system similar to the oxidation ditch but employing Simcar, vertical spindle aerators. These enable the depth to be greater than with TNO rotors. In passing once round the complete circuit the liquor goes through two cycles in which intense aeration near the rotors is followed by falling DO concentration, to the extent that parts of the circuit are anoxic (see p. 235) and de-nitrification can occur.

The systems described above are those which are currently most common in the U.K. Coarse bubble air diffusion systems have been used to a small extent in addition. While there is much evidence that fine bubbles enable a higher proportion of the oxygen in the air to be transferred to the liquor, coarse bubble systems can be operated under lower pressure though with a higher flow. Fan type blowers can be used near each tank rather than high-pressure machines in a separate building. Air filters are not needed.

A much wider variety of aeration equipment both air-bubble and surface aeration is available in the U.S.A.

The I.C.I. deep-shaft system and oxygen-activated sludge systems have their own special characteristics and are described in Sections 11.10 and 11.11.

11.9 Mass transfer of oxygen

The micro-organisms in activated sludge use the oxygen in solution for respiration. The objective of aeration devices (in addition to mixing and keeping sludge in suspension) is to dissolve air continuously into the liquor so that an adequate steady concentration of DO is maintained.

The mass transfer (or mass flow) of oxygen into solution depends on the oxygen deficit $(C_s - C)$, in which C_s is the DO for complete saturation and C is the actual DO of the solution. If C is changing with time the rate of transfer of oxygen will also be changing and the rate of change of DO must be expressed in differential form:

$$\frac{dC}{dt} = K_L a (C_s - C) \tag{11.1}$$

The coefficient K_L is a measure of the rate of diffusion of oxygen through unit area of surface into unit volume of solution for unit DO deficit. In an aeration tank new liquid surface into which oxygen can diffuse is being created continuously, indeed this is the main purpose of the aerator. Consequently it is not practicable to treat the surface area a as a separate variable so $K_L a$ is treated as an entity. It can be measured by carrying out a standardised test on an aerator and then used to find the number of aerators needed for a given application. These aspects of the application of equation (11.1) will be treated in turn.

To evaluate aerator performance, tests are carried out in a tank with clean water from which dissolved oxygen has been removed by treating with sulphite together with a cobalt salt which acts as a catalyst. DO in the tank is measured at intervals of time after starting the aerator by sampling or by use of DO probes. Concentrations are measured at several points in the tank. The test is run until DO reaches near-saturation level.

In such a test, C is varying with time so it is necessary to use the integrated form of (11.1):

$$K_L at = - \ln(C_s - C) + \text{constant}$$

A graph of $\ln (C_s - C)$ against t using the values of C at successive times gives a straight line of slope $- K_L a$. However, K_L and C_s depend upon temperature and C_s depends upon atmosphere pressure so the result is corrected to a standard temperature of 10 °C or 20 °C and to a standard pressure of 760 mm of mercury. The corrections can be made by using:

$$(K_L a)_{t_1} = (K_L a)_{t_2} (1.024)^{t_1 - t_2} \tag{11.2}$$

and

$$(C_s')_p = (C_s)_{760} \left\{ \frac{P - p}{760 - p} \right\} \tag{11.3}$$

in which p is the saturated water vapour pressure in mm of mercury.

However, the temperature correction is commonly neglected as effects of temperature changes on $K_L a$ and C_s tend to compensate for each other.

Considering the aerator in normal use a steady state exists, the oxygen flowing into the system being used by the sludge organisms and C having an equilibrium value with dC/dt constant. The flow of oxygen in which we are interested is given by $V. dC/dt$ or, using (11.1), by:

$$OC = K_L a(C_s - C)V \tag{11.4}$$

in which OC is the oxygenation capacity of the aerator and V is the volume of the tank in which it is operating. However, both $K_L a$ and C_s are different for mixed liquor than for clear water so factors α and β have to be introduced as follows:

$$OC = \alpha K_L a(\beta C_s - C)V \tag{11.5}$$

Fig. 11.10 Simcar aeration plant at East Calder Works, Midlothian (*Courtesy Simon-Hartley, Stoke-on-Trent, Ltd*)

in which $K_L a$ and C_s are the clean water values. The factor α is less than unity if soaps, etc., are present but it seldom less than 0.85 for sewage, β is a little less than unity (0.9 to 0.95) for sewage. If a prediction of OC is needed at temperature and pressure other than standard, equations (11.2) and (11.3) are used to convert C_s, the temperature correction usually being neglected for the reason given above.

Rather than quote $K_L a$ values equipment manufacturers of surface aerators commonly give values of OC per kW of shaft power at standard temperature and pressure in clean water at zero DO for a stated volume of tank. Thus the ratio of the actual working OC per kW to the quoted value (in a tank of the stated volume) is:

$$\alpha \left\{ \left(\frac{P - p}{760 - p} \right) \beta - \frac{C}{C_s} \right\}$$

Performance of an aerating device depends on the shape as well as the volume of the tank in which it operates. Consequently, it is important that tests are carried out in a tank similar to that in which the aerator is to operate.

With bubble-aeration systems an important factor is the proportion of oxygen in the air which is transferred to the liquor as the bubbles rise from the floor to the surface of the tank. This proportion increases approximately linearly with the depth of the tank. For the fine bubbles of the diffused-air system, about 30% of the oxygen is transferred from a flow of 0.25 l/s per dome at a depth of 4.5 m in clear water. Transfer is reduced by impurities in the liquor by a factor which increases along a plug-flow tank as the liquor becomes weaker. More power is needed per unit volume of air as tank depth increases, since there is more hydrostatic pressure to overcome, but power per unit weight of oxygen transferred is virtually independent of depth because of the improvement of transfer with increase in depth. Consequently diffused air has an advantage over surface aeration where limitations of site area dictate the use of deep tanks.

11.10 The I.C.I. deep-shaft system[1]

In this system a lined borehole or shaft sunk into the ground is used in place of an aeration tank. The shaft is 50 to 150 m deep and 0.5 to 10 m in diameter. A tube is fitted into the shaft. Mixed liquor circulates continuously passing upwards through the outer annular space and downwards through the tube. Liquor circulates at a much higher rate than that of the throughput, the shaft operating as a completely mixed system. Velocity in the shaft is of the order of 1 to 2 m/s.

To start up the circulation, air is injected at a depth of several metres in the annular upcomer. The air bubbles reduce the density of the mixture above the level of injection so that it is less than that in the downcomer at the corresponding level causing circulation to commence. Once circulation has become established air is injected in the downcomer at a depth of 20 to 40 m

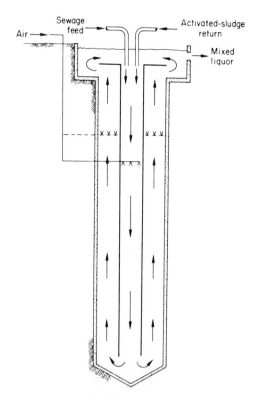

Fig. 11.11 I.C.I. Deep-shaft system (*Courtesy Paterson Candy International Ltd*)

depending on design requirements. The velocity initially established is sufficient for the bubbles injected in the downcomer to be carried downwards. As flow continues downwards hydrostatic pressure increases and the bubbles are completely dissolved. In the upcomer bubbles are formed again as hydrostatic pressure lessens. There are sufficient bubbles in the upper part of the upcomer for circulation to be maintained without further injection in the upcomer. The upcoming bubbles contain carbon dioxide resulting from respiration as well as residual oxygen and nitrogen. The gases are released from the free surface at the top of the shaft which is widened to form a 'disengagement chamber'.

The deep-shaft principle results in highly effective oxygen transfer in several ways. The turbulent nature of the flow promotes a high rate of bubble surface renewal. Pressure at the bottom of the shaft is 5 to 10 atmospheres, giving a corresponding increase in oxygen solubility. Bubble contact time is 3 to 5 minutes which is much longer than in conventional bubble aeration systems. Since the bubbles are rapidly carried away high rates of air injection can be used.

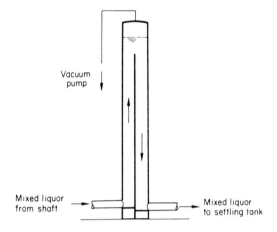

Fig. 11.12   Vacuum degassing column for I.C.I. Deep-shaft system (*Courtesy Paterson Candy International Ltd*)

Mass of air injected per unit of energy is high and depends on the depth of injection. The range is 3 to 5 kg oxygen per kW h.

Lower volumetric capacity is needed than for conventional aeration tanks, a detention period of 1.5 h having been found adequate for domestic sewage. Primary sedimentation is unnecessary.

Sludge floc from the deep-shaft process needs to be degassed under vacuum before it will settle readily so a column (Figure 11.12) has to be interposed between the aeration shaft and the final settling tank for this purpose.

Surplus sludge production is somewhat less than in conventional aeration and this is believed to be due to the balance between micro-organism energy generation and utilisation being struck at a different point so that a higher proportion of carbon is converted to carbon dioxide.

Being a mixed system deep-shaft treatment has a good resistance to shock loading. Possibilities of sludge bulking are claimed to be reduced.

Though calling for specialised techniques and equipment the sinking of boreholes and shafts is of course well established in the mining and other industries.

The development of the deep-shaft system was a by-product of research into aerobic fermentation of methanol for single-cell protein production. Its application to the treatment of waste waters of high BOD exploits its potential more fully than is the case with domestic sewage.

11.11 Oxygen activated sludge systems[2]

The use of oxygen in place of air offers the possibility of maintaining a given DO level in mixed liquor with a smaller expenditure of power upon gas transfer. Since air contains about 20% of oxygen the partial pressure of oxygen in air is only about one-fifth of that in a pure oxygen atmosphere, and it is to be expected therefore that the oxygen transfer coefficient $K_L a$ will be about 5 times as great for oxygen as for air at given temperature and pressure.

Though the energy for oxygen transfer may be reduced by using oxygen in place of air, energy is needed to produce oxygen by separating it from air.

There are two processes in use for oxygen production. In the older cryogenic or Linde process adiabatic expansion is used to cool and liquify air which is then fractionally distilled to produce oxygen of about 99.7% purity. The more recent Skarstrom or PSA (pressure swing absorption) process employs a zeolite which absorbs nitrogen from air at a pressure of 200 to 300 kPa (2 to 3 atmospheres). Nitrogen is subsequently desorbed from the zeolite by reducing the pressure to atmospheric. PSA produces oxygen enriched air of 79 to 95% oxygen depending on operating conditions.

Cryogenic plants of 1500 tonne oxygen per day output are typical. PSA plants are currently available in the 1 to 50 t/d range. Up to 20 t/d they are claimed to be lower in capital cost than cryogenic plants but require more power for operation. Economies of scale apply to capital costs but not significantly to energy requirements. At present a degree of unreliability in PSA plants must be envisaged due to failures of valves (which typically operate on a 2-minute cycle) so standby oxygen storage is required.

Energy requirements are about 0.35 kW/kg for cryogenic oxygen production and 0.35 to 0.6 kW h/kg for PSA.

Overall energy requirements for oxygen applied to activated sludge treatment are in the range 0.65 to 1.0 kW h/kg for maintenance of zero DO concentration compared with 0.5 to 0.7 kW h/kg for air application. However, to maintain DO at 5 mg/l the energy input to an air plant would have to be doubled while that to an oxygen plant would be increased by only about 5 per cent. Thus high DO levels can be maintained in an oxygen plant for no greater total energy input than for an air plant.

The UNOX system is shown diagrammatically in Fig. 11.13. It was developed by the Union Carbide Corporation in the U.S.A. and is marketed in the U.K. by Wimpey Unox Ltd.

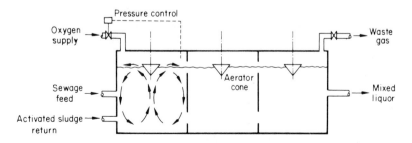

Fig. 11.13 UNOX oxygen activated sludge plant (*Wimpey-Unox Ltd*)

The plant consists of a number of surface aeration tanks operating in series. In order to prevent the loss of oxygen to the atmosphere the tanks are covered, oxygen being supplied under pressure. Controls maintain a constant gauge pressure of 20 to 50 mm of water in the gas space. The proportion of oxygen in the gas space becomes progressively lower in the series of compartments due to the carbon dioxide produced by oxidation and the accumulation of nitrogen initially present in solution. The exhaust gas stream is about one-fifth of the input gas stream and has an oxygen content of 40 to 50%. Typically 85 to 90% of the input oxygen is utilised.

Pilot-scale studies have been carried out using Simplex aerators in covered tanks supplied with oxygen.

Variants of diffused air systems include those producing bubbles less than 0.2 mm so that they are completely dissolved before reaching the surface (FMC Corporation, U.S.A.) and those using conventional bubble aerators with recycling of gases from the enclosed space above the tank to achieve higher utilisation of the oxygen supplied. A third means of increasing oxygen utilisation is to inject the bubbles into a downward-flowing stream of liquor so that their contact time is increased.

The Vitox system of the British Oxygen Company employs a sidestream injector technique. This has been used effectively to improve performance of existing surface aeration tanks when DO levels were too low in some parts.

As a result of many trials at both pilot- and larger-scale installations a number of benefits from the use of oxygen are now widely accepted though these benefits have not always been found in practice, suggesting that there is still much to be learned about the most effective way of operating an oxygen plant in given circumstances.

The benefits may be summarised.

1 Higher F/M ratios can be used than for air plants and the volume of aeration tank reduced. Savings in capacity are particularly significant when input BOD is high.

2 Even at higher F/M ratios less surplus sludge is produced so that sludge disposal costs are lower. Sludges from oxygen plants settle better, possibly requiring smaller final settling tanks, and consolidate better.

3 Due to the higher DO levels which can be maintained and the greater control which can be exercised, oxygen plants are able to cope with wide variations in BOD and with shock loads to produce effluents with more uniform high quality. The possibility of sludge bulking is reduced.

4 With covered tanks, malodorous waste can be treated without nuisance since the exhaust gases can be scrubbed before release.

Against these are some disadvantages.

1 Nitrification is not normally accomplished. This is thought to be due to higher concentrations of carbon dioxide lowering the mixed liquor pH, an effect which is less serious for wastes of high alkalinity. Nitrification has been achieved in a single stage plant by adding sodium hydroxide. The low sludge age in a single-stage oxygen plant may also have a bearing on the lack of nitri-

fication. If nitrification is required, second-stage oxidation possibly using an
air plant or filters might have to be envisaged.

2 Though the tanks are likely to be smaller they need to be roofed and there
is the additional cost of oxygen production plant to be considered. Another
constructional factor is that the use of mild steel has to be avoided in an
oxygen-rich atmosphere.

3 Highly skilled attendance is required particularly if the oxygen is
produced on site.

4 There are hazards to safety particularly in increased fire risk. UNOX
plants have equipment for detecting combustible gas and automatically
purging the tanks with air after shutting off the oxygen supply. This is done
when the gas mixture exceeds 25% of the explosive limit of propane.
Aerators are shut down if the mixture exceeds 50% of that limit.

From the above it may be deduced that where a waste is high in BOD
and/or highly variable, where space is limited and where a relatively 'high
technology' system is acceptable the use of oxygen in place of air is likely to be
highly competitive. So far too little information and experience seem to be
available to judge whether oxygen systems will displace air systems more
widely. Furthermore, it appears to be prudent to carry out pilot studies
whenever the use of an oxygen system is to be considered in a particular
application.

11.12 Mathematical formulation of the activated sludge process

A mass balance of the completely mixed process may be expressed

$$Qb_1 + rQb_2 - Q(1 + r)b_2 = UV$$

whence
$$Q(b_1 - b_2) = UV$$

in which Q is the rate of flow, b_1 and b_2 are the concentrations of biodegradable material in the input and output respectively, U is the rate of
removal of biodegradable material per unit volume, V is the volume of the
aeration tank and r is the ratio of returned activated sludge flow to inflow.
The second term of the first equation accounts for the mass of biodegradable
material in the liquor carrying the returned sludge. Note that r disappears.
Commonly 5-day BOD is used for b_1 and b_2 though it is recognised as a less
than ideal parameter.

The rate of concentration-dependent reactions can be expressed

$$\frac{dc}{dt} = -kc^a$$

in which c is concentration at time t and k is a rate constant. Reactions are
termed first-order, second-order etc., corresponding to $\alpha = 1,2$ etc. A
zero-order reaction is independent of concentration. The removal of

biodegradable material in the activated-sludge process is commonly considered to be psuedo first-order. For a completely mixed system the mean concentration in the aeration tank is that of the effluent so (assuming first-order kinetics) we can write:

$$U = kb_2$$

and from above

$$b_1 - b_2 = UT$$

since $T = V/Q$ the mean detention time in the aeration tank.
Hence

$$\frac{b_2}{b_1} = \frac{1}{1 + kT}$$

Though the mean time for a single pass of the aeration tank is $T/(1 + r)$, it can be shown that for recirculating processes, because decreasing fractions of flow experience increasing numbers of passes, the mean residence time is independent of recirculation rate. Consequently the mean residence time is equal to that for zero r and is therefore equal to V/Q.

The following extension to plug-flow systems is tentatively offered.
Consider first a series of N completely mixed units each of volume V/N. For each unit the residence time is $T/N(r + 1)$ and for ith unit,

$$\frac{b_i}{b_{i+1}} = \frac{1}{1 + kT/N(r+1)}$$

For the series,

$$\frac{b_N}{b_1} = \frac{b_N}{b_{N-1}} \cdots \frac{b_3}{b_2}\frac{b_2}{b_1} = \left\{ \frac{1}{1 + kT/N(r+1)} \right\}^N$$

assuming the same k or some appropriate mean k can be used for each unit.
A true plug-flow tank may be considered as $N = \infty$ for which

$$\frac{b_N}{b_1} = \exp\left(- \frac{kT}{r+1} \right)$$

In these expressions the dilution of the inflow by the return must be taken into account by means of a mass balance at inlet:

$$b_1(1 + r) = b_0 + r b_N$$

in which b_0 is the concentration of the waste to be treated. Analogous but more complicated balances are conceivable for incremental feed.

For a batch process integration of $db/dt = - kb$ gives $b = b_1 \exp(- kt)$, the concentration after time t with initial concentration b_1. This is analogous to the exponential expression above, when account is taken of the dilution of the sample with sludge suspension at the beginning of the test.

Though it is possible to compare mixed and plug-flow systems on the basis of the various equations the validity of conclusions drawn using the same value of k is doubtful.

An interpretation of k may be developed from Monod's equation

$$\mu = \mu_{max} \frac{S}{K + S}$$

which relates μ, the specific growth rate of microorganisms to S the concentration of the soluble growth-limiting substrate on which they depend for nutrient. The specific growth rate (rate of increase of mass per unit mass) has a limiting value μ_{max} to which it is asymptotic as S increases. The 'saturation constant', K is introduced to formulate this asymptotic property and is the value of S for which μ is $\frac{1}{2}\mu_{max}$.

If X is the mass of micro-organisms at time t,

$$\frac{dX}{dt} = \mu_{max} \frac{XS}{K + S}$$

and

$$\frac{dS}{dt} = \frac{-\mu_{max} X}{Y} \cdot \frac{S}{K + S}$$

where $Y = -dX/dS$, the mass of micro-organisms produced by the metabolism of unit mass of substrate.

In terms of activated sludge (12) may be written

$$\frac{db}{dt} = -\frac{\mu_s s_a}{Y_s} \cdot \frac{b}{K_b + b}$$

with b as BOD, s_a as the mixed liquor suspended solids, Y_s the mass of sludge growth per unit mass of BOD removed, μ_s is the maximum specific growth rate and K_b the value of b for which the growth rate is half the maximum. BOD and SS as measures of substrate and micro-organism concentrations are not ideal. Using concentrations of volatile suspended solids is preferable.

When b is very small in comparison with K_b,

$$\frac{db}{dt} = -\frac{\mu_s s_a}{Y_s K_b} \cdot b$$

and we see that kT in the equation above may be written $\mu_s s_a T/(Y_s K_b)$ for low b which is the case for medium rate and low F/M plants producing well-oxidised effluents.

Some versions of this formulation eliminate Y by introducing sludge age (or mean cell residence time) and use this as a primary parameter.

While some use has been made of mathematical formulae in designing plants for trade waste treatment, particularly in the U.S.A., they are less commonly used for municipal sewage plants where accumulated experience remains the main guide. The formulae are mainly of value in providing a basis for the treatment of experimental results from laboratory and pilot scale tests.

Biological filters

11.13 Introduction

Percolating (or trickling) filters consist of beds of stone, slag or similar frag-
mental material. Settled sewage is sprinkled on the surface of the bed and
flows thinly over the stones so that ample air voids are left. The stones
become coated with a gelatinous film containing micro-organisms which
oxidise organic material in the sewage.

In this context the term filtration must be regarded as meaning passing of
flow through a device which causes the removal of some characteristics of the
liquor. This meaning is wider than that of the mechanical straining-out of
solid particles which does not occur in percolating filters to any significant
extent. Percolating filters are sometimes known as bacteria beds to avoid the
word 'filter' but organisms other than bacteria play an important part.

As in biological oxidation by activated sludge, the organisms in percolating filter film which are primarily responsible for the removal of oxygen
demand are bacteria. Since many different organic compounds are present
in sewage and a particular species of bacteria can thrive only on a limited
range of compounds, many different species are present and the population
varies from one works to another depending upon the nature of the sewage.

As well as the heterotrophic bacteria which use complex organic compounds, there are autotrophic bacteria which oxidise simple compounds.
The autotrophs are responsible for the conversion of nitrogen from the
ammoniacal to the nitrite form and for the conversion of nitrite to nitrate
nitrogen. Since these organisms thrive best on sewage which has already
been partly oxidised, the lower parts of a percolating filter suit them better
than activated sludge where as the sludge moves cyclically through the
process conditions are continually changing.

Bacteria are by no means the only living organisms inhabiting filter beds.
Fungi live on complex organic waste products and may be in competition
with bacteria for the same food. In some circumstances, for example where
the sewage contains certain trade waste, fungi may be predominant particularly near the surface.

Protozoa which can use solid particles as food are also present in filter bed
film.

The variety of forms of life such as fungi, protozoa and algae is generally
wider in percolating filters than in activated sludge. The organisms mentioned so far are present in the film, but percolating filters are inhabited also
by worms, insects, arachnids (spiders and mites) and sometimes snails.
These higher forms of life feed upon the film and therefore indirectly on
nutrients from the sewage since the micro-organisms of the film use
the nutrients partly to make new cell material. In feeding upon the film, the
'grazing' fauna keep its growth in check and maintain a satisfactory balance.
Apart from this, their excretory products are in some cases readily oxidisable

by the lower organisms so that they form an additional link in the purification process.

The population of macro-fauna depends on the nature of the sewage, being less varied where the sewage is strong. The rate, frequency and method of application of sewage to the surface also influence the population. Thus, the application of flow by spaced jets may be advantageous in that alternate wet and relatively dry zones are created near the surface of the bed in each of which different creatures are encouraged.[3]

The thickness of the biological film coating the stones varies seasonally, tending to be greater in winter than in summer. Surplus film detached from the stones is carried out in the effluent and is known as humus. It is settled out in humus tanks and often disposed of with primary sludge. In spring a good deal of film is released from the filter. It is probable that macro-fauna play a part in causing film to be detached but whether the causes of the high degree of release in spring are the same as those of the continual release is not known. In high-rate filtration as practised in the U.S.A., the flow of sewage may be more important in controlling film thickness, by washing out surplus, than the macro-fauna.

In some circumstances, film accumulation is so extensive, particularly in winter, as to block the surface of the bed and to cause ponding. This is an important factor in limiting the flow of a particular sewage which can be successfully treated on unit volume of filter. Reducing sewage strength and increasing rate of flow will often enable ponding to be avoided. Consequently the dilution of raw sewage with a proportion of the filter effluent often enables net rates of sewage throughput to be increased. Alternating double filtration described later is another expedient. Reducing the impurity load by removing fine suspended matter with activated sludge (bioflocculation) has been mentioned above, and also enables rates of throughput to be increased.

Raking the surface of a ponded filter gives temporary relief but the provision of more capacity or conversion to another method of operation is the only permanent cure. The root cause of ponding is biological, being connected with the balance of film growth and macro-fauna activity, so investigation of the ecology of a filter is of benefit.

Though insects have an important role in percolating filtration, the insect population may become so large in summer as to cause a nuisance where dwelling-houses are near to the works. Insecticides have to be used with great care to avoid adverse effects on treatment through changes in the ecology, and also to avoid concentrations in the effluent which would be toxic to fish.

Small numbers of the spores of the organisms which inhabit filter beds are present in sewage or in the environment and do not have to be deliberately introduced when a new filter is started. The bed is a suitable habitat for these organisms and they multiply during the first few weeks of operation. There is an advantage in starting up new filter beds in summer. The film is established first, followed by the macro-fauna. During the time when the bed is being brought to maturity, the impurity load should be increased gradually to normal operating level so that the film does not build up too rapidly before grazing fauna are established. If ponding conditions are allowed to occur, the establishment of grazers will be delayed.

A detailed account of the biology of filters and of activated sludge has been given by Hawkes[3] who has done a good deal of original work in this field.

11.14 Constructional features

The stones or other fragments used in filter beds are known as media. The cost of the media is a most significant part of total cost of a filter scheme and is the main reason for the high capital cost of filtration compared with activated-sludge treatment. As many different materials can be used for filter media, the choice has often been governed by what is available locally rather than by what would be ideally suitable. Stone, from various geological formations, slag, clinker, coke and even coal have been used. Rough materials with a high proportion of surface to volume have an obvious advantage. The material must not break up during handling nor due to weathering or the action of sewage. There is a British Standard for filter media (B.S. 1438: 1948).

Since void spaces for air are essential, the fragments must be uniform in size and roughly cubical rather than long or flat. The most commonly used size is 25 to 40 mm which seems to give the best overall results. With larger sizes, the tendency for ponding in winter is reduced but results are otherwise inferior. The lowest stratum of media in the bed is made from quite large stones, often 100 to 150 mm, to restrict the passage of the small media to·the underdrains and to allow air to penetrate more easily.

Filters are most commonly 1.83 m (6 ft) deep in Britain but vary more widely elsewhere. Treatment capacity is governed mainly by volume so in some cases head loss is saved by using shallow filters of about 1.2 m depth at the expense of occupying a larger area, and conversely deeper filters have been used where area is restricted. It now appears that depth has a small effect on treatment capacity, a deep filter being slightly better than a shallow one of the same volume.

In order to collect liquid flowing through a filter, a concrete floor is provided. The weight of media in a filter of normal depth can safely be carried by the ground on most sites so the floor needs only nominal reinforcement. However, the reinforcement needs to run in two directions at right-angles and to be placed both top and bottom as a precaution against bending moments of either sense where local soft spots are spanned.

Half-round tiles are laid upon the floor to support the media and to allow free passage of effluent and air. The floor is laid to fall at a gradient of about 1 in 100 to 200. In the case of rectangular beds, a collecting channel runs along one side of the bed. Circular beds may have a circumferential channel with the floor falling radially outwards, or a channel running half-way round, with a plane fall towards the channel. Another expedient is to have one or more internal channels with the floor sloping at right-angles to them.

Walls to retain the media are most commonly of brickwork. Large pieces of slag have been used on some works to build rough walls. Planning authorities have been known to require the use of material which resembles local stone, and tinted rough-cast concrete slabs have been used.

Wherever possible, filters are built with their bases at ground level so that the underdrains have free access to the air. Walls are sometimes perforated but this may benefit no more than a small part of the volume and is no substitute for free flow of air between the floor and the surface. Where a filter has to be built partly below ground level, vertical pipes are sometimes provided in the bed to form an air passage to the underdrain system. Galvanised wire frameworks have been used to provide an air passage by separating the media from the wall.

Circular filter beds with rotating sprinklers, Fig. 11.14, are most common. The sprinkler consists of a central column carrying four perforated pipes which are 100 to 200 mm above the surface of the media. Wire stays are used to support the pipes from the head of the column and to brace the pipes in the horizontal plane.

Each pipe has a line of holes drilled along one side and the jet reaction of the sewage issuing from the holes serves to rotate the sprinkler. The spacing of the holes is wider near the centre of the bed than at the periphery so that the rate of application of sewage to unit area of the bed will be uniform. In some models, the holes are provided with replaceable bushes. Specially shaped bushes and fish-tails are sometimes used to spread the jets, but a fairly localised jet may be better for the development of varied macro-fauna than uniformly spread flow.

The outer ends of the sprinkler pipes are fitted with discs which are normally closed but can be opened periodically to flush out deposits.

The detailed design of the centre column differs a good deal between models. In many cases a large proportion of the weight of the rotating part is supported by floating it on a double seal of liquid trapped in annular spaces between the fixed and rotating parts. The seals have to be quite deep where water is used but more compact arrangements are possible with mercury seals.

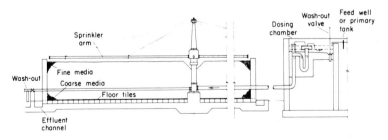

Fig. 11.14 Percolating filter and dosing chamber (*Based on equipment of Adams-Hydraulics Ltd*)

Some models have centre columns which can be buried in the media while others require a wall of brickwork to be provided to form a space round the column and to give access for maintenance.

The pipe feeding the centre column runs through the media rather than under the bed and is continued to the far side of the bed where a valve is provided for flushing and to give access for rodding. This washout pipe should preferably discharge to the works drainage system so that flow through it can be pumped back to the works inlet. There would be little harm for wash-out purposes in discharging the pipe to the effluent-collecting channel but there would be a risk of a careless operator allowing the filter to be by-passed continuously.

In order to ensure that the rate of flow fed to the distributors is always sufficient to cause rotation, a dosing chamber fitted with a bell siphon is used. Flow to the distributor occurs only when the siphon is primed and the levels are such that there is sufficient head to rotate the distributor. When the flow to the dosing chamber is less than the discharging capacity of the siphon, water level in the chamber falls and siphonic action ceases. No more flow passes to the distributor until the water level has risen far enough to reprime the siphon. Where distributors are fed from fixed output pumps, dosing chambers can be eliminated, the balancing storage being provided by the pump-well.

Circular filter beds are often arranged in groups of four around a common feed-well housing weir penstocks and dosing chambers. The four feed pipes radiate from the well.

Distributors for circular beds range in size from 3 to 37 m in diameter. The smallest sizes are driven by some form of paddle wheel rather than by reaction of the jets, and dosing is sometimes via a pivoted bucket designed to tip when it is full.

On large filter installations, long rectangular beds are preferred as they use the space available more fully. Rectangular beds are divided longitudinally by a feed channel whose walls carry rails on which the distributor runs. Sewage is siphoned from the channel by a pipe attached to the distributor. Before passing to the filter arms, the flow is used to power a simple paddle-wheel turbine which drives the wheels of the distributor. Flow can be directed to either side of the turbine by a two-way valve. This is actuated by a lever which is pushed over by a stop at each end of the bed to reverse the direction of motion. Distributors hauled by a cable device from stationary electric motors are used if insufficient head is available.

Distributors are available for rectangular beds up to 12.2 m wide on each side of the centre channel.

The beds need not necessarily be rectangular. At Minworth, Birmingham, one of the largest filter installations ($17\frac{1}{2}$ ha), the beds are of parallelogram form, the distributors being set at an angle to the centre channel.

In planning filter beds with distributors travelling backwards and forwards it is necessary to decide whether the distributors are to discharge when travelling in both directions or in one direction only. In the former case, the frequency of dosing a particular area of the bed will depend on the distance from the end. At the half-way point, the media will be dosed twice per cycle at regular intervals while at the extreme ends dosing will be once per cycle. Elsewhere dosing will be twice per cycle but at unequal intervals, the inequality increasing towards the ends. Where distributors are designed to discharge when travelling in one direction only, the frequency of dosing is uniformly once per cycle. To provide for a steady flow of liquor, arms on one side of the centre channel discharge when the distributor is travelling one way and those on the other side discharge when the direction of travel is reversed. The effect is the same as if the sprinkler arm rotated on plan at the ends and the action is analogous to that of a distributor on a circular bed.

Hawkes concluded that frequency of dosing is an important factor in controlling film growth in winter and as a consequence the reconstructed Minworth filters are arranged to discharge when travelling in one direction only and provision is made for varying the speed of travel and hence the dosing frequency.

If it is desired to employ very long rectangular beds, without unduly increasing the dosing frequency, more than one distributor can be provided for each bed. The advantages of shorter beds, each with its own distributor (flexibility of operation and ease of access), should not be overlooked.

11.15 Special forms of filter

Enclosed filters have occasionally been constructed. These enable temperature variations to be reduced and avoid odour and insect nuisances. Forced draught is needed and is more conveniently applied in the downward direction. For economy of construction, an enclosed filter needs to be deeper than the normal 1.8 m to avoid large areas of roofing. Depths adopted vary from 3.5 to 5.5 m. Enclosed filters have given good results enabling the treatment capacity per unit volume to be increased by factors of two or three. It is possible that enclosed filters would have a useful application for the treatment of trade waste at source.

Plastics manufacturers have developed a number of products for use as biological filter media. Some consist of small rings or tubes forming random packings, others of corrugated and perforated sheet assembled into units which can be stacked together. The proportion of voids in filters packed with these materials is about 95 %, double that of mineral media, and the internal surface area per unit volume is 2 to 3 times that of mineral media. Because of the larger proportion of internal surface, the flow per unit plan area of filter (irrigation velocity) has to be high to keep the surface sufficiently wet. Filters with plastics media tend therefore to be made deeper than those with mineral media. Since the material is light, depths up to 8 m are feasible, saving site area but incurring energy costs for pumping.

Plastics media filters offer economical means of partial treatment for wastes of high BOD and have a role in the treatment or pre-treatment of trade wastes. They have been used at sewage works to relieve overloaded mineral media filters by pre-treatment of part of the flow.

It has been found feasible to use random packed plastics media filters without prior sedimentation treatment provided finer screens (1.5 mm spaces) are used[4]. The irrigation velocity was 0.25 m/hr. Total sludge production was less than would have been the case with conventional filter treatment, and the sludge was more easily dewatered. The capacity of an existing works could be increased by installing a filter on this principle upstream of the primary sedimentation tanks which could then serve as humus tanks.

11.16 Alternating double filtration

In this system of operation the filters are divided into two groups through which the sewage passes in series. At intervals of a few days the duties of the groups are exchanged. The principle is shown in Fig. 11.15.

The total volume of bed needed to treat a given flow on alternating double filtration is about half that needed for single filtration. If double filtration is used up to the maximum rate of flow, all flow has to be pumped between stages and the total humus tank capacity is double that for single filtration as humus tank treatment is given twice. It is sometimes feasible to change over to single filtration whenever the flow exceeds a given value. This saves

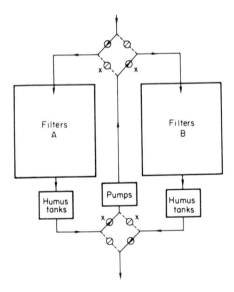

Fig. 11.15 Alternating double filtration. Operation alternates between the circuit shown and its mirror-image. (Recirculation could be operated by opening all valves. For straight-run operation, valves X would be closed, other valves remaining open.)

pumping costs and, where the change to single filtration is made whenever the flow exceeds half the maximum, the humus tank capacity need be no more than for a plant operating continuously on single filtration.

Alternating double filtration enables a higher rate of treatment to be achieved because it provides a ready means of controlling film growth. The possibility of doing this arose from the observation by Whitehead and O'Shaughnessy that partially treated sewage was capable of restoring a ponded filter to a normal condition which followed from the application of bioflocculation as a pre-treatment process at Birmingham. When filters are operated alternately as primary and secondary treatment units, the film alternately grows and disintegrates. With rates of application as high as those used in alternating double filtration, single-pass filters would pond owing to the high rate of supply of nutrients and the discouragement of macro-fauna. Exchanging the roles of the filters when the film on the primary treatment unit is becoming excessive enables ponding to be avoided.

Distributors for alternating double filtration need more capacity than those for single filtration though fewer will be needed as the plant is smaller. The rate of application to unit surface area at any given time is increased by a factor of about four since the overall depth of filtration is doubled and the rate of application to unit volume is also roughly doubled. However, if double filtration is restricted to flows not greater than half the maximum flow rate, distribution capacity need only be greater than that for single filtration by the factor of increased treatment capacity per unit volume.

11.17 Recirculation

Pumping back a proportion of the treated effluent and using it to dilute the sewage fed to filters enables results to be achieved comparable with those of alternating double filtration.

In alternating double filtration, a particular filter is subjected to high flows of untreated sewage alternating with flows of partially treated sewage. With recirculation, the filter continuously receives a comparable rate of flow consisting of a mixture of treated and untreated sewage.

In the treatment of extremely strong trade waste, very high ratios of recirculation have sometimes been necessary to obtain sufficient dilution of the waste but in sewage treatment the average amount of effluent recirculated is usually about equal to the incoming flow. The rate of flow through the filter is then about twice the rate of flow of incoming sewage.

If the flow of recirculated effluent is to be in constant relationship to the incoming flow, total capacity of pumping plant needs to be quite high and arrangements for enabling a wide range of flows to be returned add further to the expense. The distributors would have to be capable of handling 6 × DWF and, as rates of application in terms of untreated sewage can be at least twice those for single filtration, the distributors need about four times the carrying capacity of single distributors. The humus tanks have to deal with flows up to 6 × DWF also. The distributors and humus tanks need to have capacities similar to those when alternating double filtration is applied up to the maximum rates of flow. Installed pumping capacity for 3 × DWF is needed.

As with alternating double filtration economies can be made in the provisions for pumping.

The simplest expedient is to recirculate at a constant rate. This rate might be equal to DWF or 20% to 30% higher if the object is to achieve an approximate 1 to 1 ratio averaged out over wet and dry conditions. In these cases, humus tanks have to treat flows up to 4 or 5 × DWF and distributor carrying capacity is about 3 times that for single filtration. Installed pumping capacity amounts to 1 to 1.3 DWF and variable rate pumping is avoided.

Another alternative is to arrange for the sum of sewage and recirculant flows not to exceed some chosen limit. This may be done merely by restricting recirculation whenever the flow is high or by arranging for the sum of inflow and recirculation to be more or less constant, giving high ratios of recirculation at low rates of inflow and low rates of recirculation at high flows. If 3 × DWF is chosen as the maximum rate of total filter throughput, as is sometimes done, recirculation is zero at maximum inflow, 2 to 1 at 1 × DWF inflow and, if the principle is applied fully, about 10 to 1 at minimum input. In this case, humus tanks need be no larger than for single filtration and distributor carrying capacity is larger than for single filtration only in proportion to the greater treatment capacity achieved (i.e. it is approximately doubled). Installed pumping capacity, however, approaches 3 × DWF and provision for variable-rate pumping is needed to the same extent as in schemes where a constant recirculation rate is adopted.

In comparative tests, Lumb and Eastwood[5] experienced serious ponding in a filter operated with variable recirculation to give a nearly constant rate of total throughput. Much less ponding occurred on a bed operated with constant-rate recirculation, while on a bed operated with recirculation in constant ratio to inflow ponding was very slight. The circumstances of the tests were not fully comparable but enabled the conclusions to be drawn that variable-rate recirculation with constant throughput is hazardous for sewage from a combined system where prolonged high flows may be experienced, though it might be satisfactory for sewage from a separate system. The suspicion arises that arranging for an alternating double filtration plant to work on single filtration at high flows might be a hazardous practice for combined sewage.

An important feature of filtration with recirculation is that, though BOD removal comparable with that in single-pass filters can be achieved with rather less than half the volume of media, nitrogen is converted to nitrate form to a much lesser extent. It appears that in recirculation a greater proportion of the bed depth is used for the oxidation of carbonaceous matter and the nitrifying bacteria are restricted to a smaller proportion of the bed.

In the U.S.A., several proprietary systems of high-rate filtration have been used employing recirculation in various ways. These include filtration in stages with relatively complex patterns of recirculation and the return of filter effluent carrying humus to primary settling tanks as well as return of clarified effluent directly to the filters. Rates of application of mixed flow to the filters are often much higher than in British practice but the sewages are often weaker and the effluent standards less onerous. At very high rates of application, the scouring action of the flow is probably an important factor in limiting film growth and avoiding ponding.

11.18 Capacity of filters

In principle the *F/M* ratio discussed in relation to activated sludge also applies to filtration though it is of less service as a design parameter because there is less opportunity either to control or to assess the amount of biomass. The biomass arrives at an equilibrium naturally, growth being compensated by macro faunal activity and loss in the effluent. There is no convenient way of increasing or decreasing the biomass corresponding to the deliberate wasting of greater or lesser amounts of activated sludge.

In a filter the biomass is attached to the surface of the media so there is no doubt that the surface of media is a significant factor. However, the thickness and nature of the biological film changes with depth, biomass tending to be more concentrated near the top, particularly in single-pass filters, while the benefits of double filtration may arise in part through a more uniform spreading of the film over the depth.

A simple performance relationship[6] for filters is

$$\frac{b_e}{b_i} = \exp\left(-kS/H\right) \tag{11.6}$$

in which b_i is the BOD of the sewage applied to the filter, b_e is the BOD of unmetabolised organics left in the filter effluent, S is the specific surface of the media (m^2 of surface per m^3 of volume), H is the hydraulic loading (m^3/d of flow per m^3 of media), and k is a rate factor (m/d). In this equation S/H is m^2 of surface per m^3/d of flow so if biomass is considered to be proportional to surface area, S/H becomes proportional to biomass divided by flow which is closely analogous to the expression Vs_a/Q (which equals Ts_a) used for the activated sludge process. Thus (11.6) is analogous to equations used for plug-flow activated-sludge systems.

In fitting operational data it has been found that it is necessary to apply an exponent less than unity to H in equations like (11.6). Associated with this is the finding that time of flow through a filter varies with H. A factor of $1.08^{(T-15)}$ can be applied to the constant k to allow for a temperature T °C differing from 15 °C.

Mathematical models based on microbial kinetics have been evolved as well as empirical relationships like those above. While such studies offer a more detailed understanding of aspects of the processes, and have been helpful in guiding pilot studies, they have not been widely adopted for routine design.

Several parameters differing from each other, but inter-related have formed the basis of various design recommendations which have been used.

Mineral media filters, as commonly employed for the treatment of municipal sewage, need to produce an acceptable effluent throughout the year. During winter months in temperate climates, where most of the experience has been acquired, there will be a risk of ponding due to reduction in macro-faunal activity if the rate of application of organic material to the top of the filter is too great. It would be logical therefore to use a safe value of kg BOD/day per unit plan area of filter to determine the area of filters needed. The volume and hence the depth could then be chosen to comply with a safe value of kg BOD/day per unit area of media surfaces, coupled with the surface per unit volume (specific surface) of the media.

Hydraulic parameters are probably relevant to some extent and include flow per unit plan area (irrigation velocity), flow per unit media surface and flow per unit volume of bed. These parameters are related to each other by the depth of the bed and the specific surface of the media, and are related to the corresponding BOD loading parameters by the mean concentration of BOD of the applied settled sewage.

Since in practice depth of bed, specific surface of media, concentration of applied BOD and effluent standard lie within relatively narrow ranges there is little to choose between one parameter and another.

For single-pass mineral media filters treating sewage from mainly domestic sources to an effluent standard of 20 mg/l, typically adopted maximum loadings are 0.1 kg BOD/m³ of filter per day and 0.3 to 0.6 m³ DWF/m³ of filter per day. These loadings can be at least doubled for filters operating by re-circulation or alternating double filtration. Pilot-scale studies might well show that the loadings could be increased more than two-fold in favorable cases, though little nitrification would be achieved compared with single-pass filters.

Mineral media filters in the U.K. are mostly 1.83 m (6 feet) deep. A greater depth would require more fall through the works and perhaps a thicker base slab while a smaller depth would increase the area as well as the cost of distributors. Possibly the 1.83 m depth is an intuitive optimum but the parameter values used in the U.K. are derived from experience with this depth and, to the extent that they relate to phenomena such as winter ponding, changes might be needed in any of them for different depths of filter.

Plastics media (Section 11.15), having much higher void ratios as well as more specific surface, run no risk of ponding but in contrast require sufficient flow per unit plan area (irrigation velocity) to keep the biomass wet. Consequently with this type of filter the area must be smaller and the depth greater than for mineral media filters. Recirculation offers a means of achieving an adequate irrigation velocity. For ordered plastic media such as ICI Flocor the minimum irrigation velocity is 1.5 m/hr.

Both BOD and hydraulic loadings can be higher for plastics media than for mineral media by factors of three or four but in making comparisons it must be remembered that plastics media have been employed mostly for partial treatment of wastes of higher BOD concentration than municipal sewage.

Humus sludge production from filters is found to average between 0.3 and 0.5 g per g of BOD removal. However, there is a seasonal variation. In the U.K. the rate may be more than double the average in spring and less than half the average in summer. Where high rate filters are employed the yield of sludge from BOD removed is likely to be nearer 0.8 g. The humus in this case is likely to have a high proportion of volatile solids and to have more resistance to de-watering.

11.19 Comparison of filtration and activated-sludge processes

(1) Fundamental differences

In activated-sludge treatment the purifying organisms move with the waste while in filters the waste flows past them. This enables specialisation to occur at different levels in the filter-bed. For a similar BOD removal, single-pass filters produce a more fully-nitrated effluent than activated-sludge plants though the latter can produce nitrated effluents from many sewages if the capacity is generous.

In activated-sludge treatment air has to be forced into contact with the liquor while in filters the liquor is split into thin rivulets. Most of the comparisons noted below stem from this difference.

From the points of view of the criteria detailed below it will be seen that alternating double filtration and recirculation stand between single filtration and medium rate activated-sludge treatment.

(2) Area occupied

Unit volume of activated-sludge aeration tanks can treat 3 to 4 unit volumes of DWF per day (i.e. 8 to 6 hours nominal detention) of medium strength sewage. At 400 l per day/m³, unit volume of single-pass filter treats 0.4 of a unit volume of DWF per day. Conventional activated-sludge plants are typically twice as deep as filters. Filters therefore occupy 15 to 20 times the area of aeration tanks. High-intensity aeration plants occupy about the same area as the conventional type since the reduction in volume is effected mainly by a reduction in depth.

It is usual to provide rather more primary sedimentation capacity for filter plants than for activated-sludge plants but, in contrast, humus tank capacity for filters is somewhat less than .final settling tank capacity for activated sludge.

The area occupied by filters can be at least halved if alternating double filtration or recirculation is employed but if either of these processes is to be fully used the humus tank capacity will be doubled. Oxidation ditches need about the same area as alternating double filters.

(3) Form and cost of construction

The form of construction of filter beds is extremely simple. Apart from the floors which may be reinforced only nominally, and small structures such as feed chambers, there is no reinforced concrete work or shuttering. Activated-sludge tanks are relatively complex reinforced concrete structures but are much smaller in volume. The volume enclosed by the walls and floors of filters has to be filled with quite expensive media and this is the main factor which makes filters more expensive in first cost.

A good deal of machinery is needed for activated-sludge treatment and requires skilled maintenance. Filter distributors are rugged and almost trouble-free, requiring skilled attention only rarely.

(4) Power needs

A single-filtration plant needs no power, provided that there is sufficient fall through the works. Even if the site is so flat that power has to be used to lift the sewage through a height equivalent to the fall from the feed chamber to the effluent channel the power is only about 0.5 to 1 kW compared with about 6 kW per Ml/d for medium rate activated-sludge treatment. The power consumption of an alternating double filtration or a recirculation plant is up to about 1 kW per Ml/d. This is roughly the power needed for sludge re-cycling in an activated-sludge plant.

(5) Fall through works

About 3 m of fall is needed between top water levels of primary sedimentation tanks and humus tanks in a filter plant. In activated-sludge treatment, 1 to 2 m can be sufficient. On a flat site, the question of which form of treatment to adopt is inseparable from considerations of power requirements.

(6) Ability to withstand varying loads and necessity for skilled supervision

There is some truth in the generalisation that filter plants are better able to withstand changes in load and need less supervision than activated-sludge plants, but the statement needs qualification. The true fundamental difference is that control can be exercised in activated-sludge treatment by varying the degree of aeration and the rate of sludge return while little deliberate control of filter plants is possible. Forking over the surface is a palliative for ponded filters, and a measure of control is possible in alternating double filtration or recirculation plants. Since control decisions have to be made in activated-sludge treatment, skilled supervision is essential, though small plants treating domestic sewage or uniform waste often operate with little more supervision than filter plants. On single-filtration treatment, there is little opportunity for the application of a high degree of skill in normal circumstances, though it is true that an experienced operator may be needed to diagnose and cure ills when they arise.

A shock load which is not severe enough to cause trouble in a filter plant may warrant a change in operation of an activated-sludge plant. On the other hand, change in operation may enable certain shock loads to be dealt with which would considerably impair filter treatment. Since small, single-unit, activated-sludge plants approach completely mixed operation they should in most circumstances be better able to withstand shock loads than single-filtration plants.

(7) Other factors

(a) The total amount of sludge to be disposed of from an activated-sludge plant is often greater than for a filter plant.
(b) There is a risk of fly and odour nuisance from filter plants but not from activated-sludge plants.

11.20 Rotating biological contactors

The most common form of this device consists of a set of discs on a rotating shaft which is just above the liquid level in a half-cylindrical trough. The discs are made of metal or plastic mesh. Biomass similar to that in filters grows on the mesh and is alternately above and below liquor level as the discs rotate. This has the effect of aerating both the liquor and the biomass on the discs.

Discs are spaced at 30 to 50 mm and are between 1.7 and 3 m in diameter. Peripheral speed is about 0.3 m/s. In some versions the trough is divided laterally so that treatment is in stages, liquor flowing thinly over the divider from one compartment to the next. Another version has the discs connected to form a helix with the advantage that humus detached from the surface is moved to the outlet end of the trough. Spheres of plastic material in a rotating drum are another alternative.

The rotary unit and trough may be used solely for bio-aeration with a preceding sedimentation or septic tank and succeeding humus tank, or the settling and oxidation facilities may be combined in a single unit, the trough being large enough to accommodate sludge and humus in separate compartments.

Trials with a BioDisc (proprietors Ames Crosta Babcock Ltd) treating settled sewage suggested 6 g BOD per day removal per square metre of disc surface to achieve a 20:30 effluent. This is several times the rate of removal per unit surface area of filter media.

The rotating biological contactor provides a neat and compact plant for the treatment of small flows particularly where little head is available. Power needs are very small. Models are available for populations up to about 5000. There are larger examples in Germany.

11.21 Choice of treatment process

From a strictly biological point of view, there is a good deal to be said in favour of filtration as opposed to activated-sludge treatment but from the point of view of cost, the latter is almost always the better choice for an entirely new works. Nevertheless, filters are still used in new construction of smaller works, even for populations up to a few tens of thousands. The reason is that they are rugged and reliable and do not need so much mechanical maintenance or power, though if these factors were judged solely on economic grounds it is doubtful whether filters would be justifiable except for the very smallest works or in places where the sewage was difficult to treat with activated sludge.

Where an existing works employing filters has to be reconstructed or enlarged, the existing filters are a valuable capital asset and there is often a good case for continuing to use filtration. Conversion to alternating double filtration or recirculation may enable capacity to be increased with only a small capital expenditure on the oxidation stage. Replacing some of the mineral media with plastic media offers another economical means of increasing capacity.

The possibility of using both forms of treatment is sometimes worth considering, particularly in remodelling existing works. If a well-nitrified effluent is demanded, high-rate filtration of activated-sludge process effluent may be a better solution than single filters alone, a generous activated-sludge plant alone or a double-filtration scheme. Short-period activated-sludge treatment in advance of filtration (see p. 267) may be advantageous in certain cases, though alternating double filtration may achieve a similar overall result.

In highly-developed countries where land is not plentiful the choice for oxidation treatment lies between filtration, single or otherwise, and ordinary activated-sludge treatment. Elsewhere, algal ponds or aerated lagoons (a form of extended aeration treatment) may be preferable to either.

11.22 Stabilisation ponds

Sewage may be effectively treated merely by storage in relatively shallow open ponds for periods ranging from a few days to a few weeks depending upon climate and the degree of treatment desired.[7]

At depths to which sunlight can penetrate, stabilisation is through the agency of aerobic bacteria and an important factor is the development of algae. During the day algae, which are simple plants, use solar energy to synthesise carbohydrates to form new cell material from carbon dioxide and water. In this process oxygen is released and is used in respiration by aerobic bacteria feeding on the nutrients in the sewage. The bacteria produce carbon dioxide and other essential compounds needed by the algae.

During hours of darkness, photosynthesis cannot occur and the algae absorb oxygen and release carbon dioxide. Consequently the concentration of dissolved oxygen (DO) decreases during the night. The diurnal changes in DO depend on the relative durations of day and night. Long days and short nights encourage algal growth and, while very high DO values may be reached towards the end of the day, the concentration may descend to zero by dawn. With short days and long nights, though daytime DO values are lower, reflecting a smaller population of algae, the night-time decrease is also less and zero concentrations may not occur. Since the sewage is the source of nutrient for the algae, the rate of loading of the pond is an important factor controlling the algal population and needs to be related to the latitude and the climate.

At depths below the surface of more than $\frac{1}{2}$ to 1 m, there is insufficient light for photosynthesis, and anaerobic processes will occur. Carbon dioxide, methane and some odorous gases are released. In the upper layers of the pond the carbon dioxide is used by the algae and the odorous gases are often oxidised to a sufficient extent for them to be little nuisance. The proportions of the depth which are aerobic and non-aerobic vary seasonally and perhaps to a small extent diurnally.

If the pond is very deep ($2\frac{1}{2}$ m or more) and has a short detention period the process will be completely anaerobic and the pond will be virtually a simply constructed septic tank. Removal of BOD is of a similar order to that in primary sedimentation.

Ponds are sometimes classified as aerobic, anaerobic or facultative (both aerobic and anaerobic or changing seasonally). When a single large pond receives unsettled sewage, it is commonly referred to as a 'lagoon' rather than a pond. There will be deposits of fermenting sludge near the inlet and, though stabilisation will occur to some degree, troubles of many kinds are to be expected. This represents the most primitive form of sewage treatment.

When used deliberately for oxidation, stabilisation ponds are usually preceded by some form of primary treatment. This may range from a small anaerobic pond or septic tank to primary settling tanks with separate sludge treatment facilities. The less elaborate forms of primary treatment are the more common since stabilisation ponds are usually employed where simplicity of construction and low capital cost are important factors.

Stabilisation ponds are formed by making shallow excavations, using the earth to form embankments around them. Provided the earth has sufficient clay content it is not normally necessary to line the pond, though lining has to be considered where ground water supplies might be polluted or where the effluent is of value for irrigation.

The shape of a stabilisation pond is mainly governed by the contours of the site. Dead pockets and short-circuiting need to be avoided, but shape is less critical than for sewage-works tanks owing to the long detention periods involved. The outlet should of course be as far as possible from the inlet. Both inlets and outlets consist of pipes laid through the embankment below water level.

Stabilisation ponds have been used extensively in the U.S.A. for small communities.[8] In the Missouri and Upper Mississippi basins, ponds are designed on a basis of (22 kg BOD/ha)/day or approximately (1.6 g BOD/m^3)/day. The loading needs to be kept low to reduce odour problems in spring. In the south and southwest, the hydraulic loading associated with such a low rate of BOD loading would be too low for pond depth to be maintained because of seepage and evaporation losses. In those areas, the basis of design is (56 kg/ha)/day. Often there are two ponds in series. Supplemental aeration (usually mechanical surface aeration) enables odour to be controlled and allows BOD loading to be increased to (224 kg/ha)/day. At still higher loadings the pond would be more properly described as an aerated lagoon since the necessary degree of aeration would be accompanied by too much turbulence for algal growth.

Watson[9] describes oxidation pond practice in Israel where the effluent is used for irrigation.

A very large installation of oxidation ponds at Auckland, New Zealand is used to treat pre-aerated and settled sewage for marine discharge. The total area of 530 ha is divided between four ponds. Three ponds work in parallel and are operated on a recirculation principle, pond water being pumped back to the pre-aeration tanks. Flow corresponding to the input of sewage passes to the fourth pond before discharge. The BOD loading of the first three ponds was estimated to be (145 kg/ha)/day.

References

1 Hemming, D.L., Ousby, J.C., Plowright, D.R. and Walker, J., 'Deep-shaft' — latest position, *Wat. Pollut. Control*, **76** (4), 441–451, 1977.

2 Boon, A.G., Technical review of oxygen in the treatment of waste water, *Wat. Pollut. Control*, **75** (2), 206–213, 1976.

3 Hawkes, H.A. and Jenkins, S.H., Biological principles in sewage purification, *J. Inst. Sew. Purif.*, Pt 3, 300–323, 1951.
   Hawkes, H.A., Ecology of activated sludge and bacteria beds, in *Waste Treatment*, P.C.G. Isaac (Ed.), Pergamon Press, Oxford, 1960.
   Hawkes, H.A., *The Ecology of Waste Water Treatment*, Pergamon Press, Oxford, 1963.

4 Hoyland, G. and Roland, D., Biological filtration of finely-screened sewage, *TR 198*, Water Research Centre, Marlow, April, 1984.

5 Lumb, C. and Eastwood, P.K., The recirculation principle in filtration of settled sewage — some notes and comments on its application, *J. Inst. Sew. Purif.*, Pt. 4, 380–398, 1958.

6 Bruce, A.M. and Boon, A.G., Aspects of high-rate biological filtration of

domestic and industrial waste waters, *Wat. Pollut. Control*, **70** (5), 487–513, 1971.
7   Gloyna, E.F., *Waste Stabilisation Ponds*, World Health Organisation, 1971.
8   Svore, J.H., Waste stabilisation pond practices in the United States, *Advances in Water Quality Improvements*, Gloyna, E.F. and Eckenfelder, W.W. (Eds), University of Texas, Austin, Texas, 1968.
9   Watson, J.L.A., Oxidation ponds and use of effluent in Israel, *Proc. Inst. Civ. Engrs*, **22**, 21–40, May 1962.
10 Collom, C.C., Construction and operation of the Manukau Sewerage Scheme, Auckland, New Zealand, *Proc. Inst. Civ. Engrs*, **27**, 703–738, April 1964.

12

Sludge treatment

12.1 Introduction

The reduction of the suspended solids content of raw sewage by sedimentation leaves sludge to be disposed of. Also, as the biological oxidation processes employ the growth phase of the organisms, further solid matter is produced in the form of humus or surplus activated sludge. These secondary sludges are often difficult to deal with by themselves and are usually mixed with primary sludge, sometimes by re-settling them in the primary tanks, before treatment.

The liquid fraction of sewage does not require to be deliberately transported from the site but is merely allowed to flow into the river. In contrast, sludge has to be disposed of by transporting it from the site, or by some process such as incineration.

Raw sludge has a moisture content of 94% to 98%. The disposal problem can be reduced by eliminating some of the water, and most sludge treatment processes have this object. The volume of sludge can be reduced to one-tenth by reducing the moisture content from 96% to 60%. If the water is removed as liquid it will contain much organic material and the BOD of the liquor will be higher than that of sewage. All liquor removed in de-watering processes must therefore be given oxidation treatment.

Sewage sludges are very difficult to de-water. The reasons for this have not been completely elucidated but it is likely that the small size of the particles and consequently of the interstices is part of the explanation. Capillary and electrostatic forces play a part and some of the water exists in the form of a gel.

Most of the treatment processes are assembled in relation to each other in Fig. 12.1.

A small degree of de-watering by consolidation is accomplished in the pockets of the sedimentation tanks. This may be supplemented by further storage in consolidation tanks which may be provided with slow stirring mechanisms to encourage the growth of particle size by flocculation. Liquor is decanted from the surface.

Sludge-drying beds are the oldest method of de-watering and are still common. The sludge is spread in the open. The more readily removed water

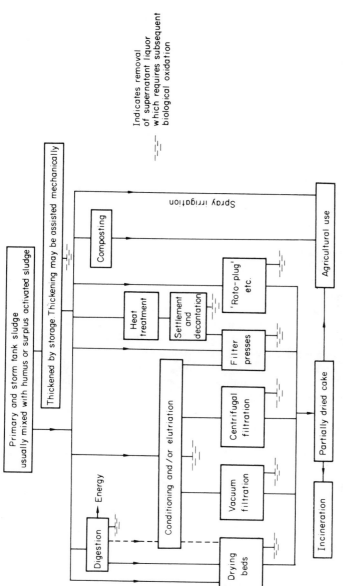

Fig. 12.1 Relationship of alternative forms of sludge treatment

is drained away and the sludge is left until sufficient water has evaporated for the surface to crack, after which the sludge can be lifted and handled as a solid material though it still contains water.

The other processes shown in Fig. 12.1 reduce or even eliminate the need for drying beds and have often been introduced where drying area has proved to be inadequate. It is not uncommon to find them working in conjunction with drying beds which are used much less than formerly owing to their high labour requirements.

Sludge digestion is an anaerobic fermentation process. It enables sludge-drying area to be reduced by as much as two-thirds and has the additional benefit of producing combustible gases which can be used for power.

Water may be removed more readily from either raw or digested sludges by causing the particles to coagulate. This may be done by adding chemical coagulants ('conditioning'). Elutriation consists of washing out small particles by an upward current of water. It can be advantageously employed before conditioning to reduce the dose of coagulant. Heating sludge at about 200 °C is another method of encouraging coagulation. After some coagulation has been achieved it becomes feasible to use vacuum filters or filter presses to force water out under pressure.

The sludge cake produced on drying beds or by one of these processes is of some agricultural value and can be used in its original state or after breaking into small fragments, sometimes with application of heat to reduce the moisture content and to destroy seeds. The cake is a low-grade fuel and can be disposed of alternatively by incineration.

Composting is an aerobic fermentation process in which sludge is mixed with vegetable matter from domestic refuse to produce material of greater agricultural value than sludge alone.

In some areas it has been found feasible to use wet sludge directly in agriculture by spraying over fields from moveable pipes or tank vehicles. The water content is of value in dry climates.

A final disposal process not shown in Fig. 12.1 is dumping at sea. For sewage works accessible to a sea-going vessel this is often the most economic method. It is usual to reduce the bulk of sludge to be transported by, for example, sludge digestion.

Some de-watering processes, such as freezing, have been shown to be possible but are not given in Fig. 12.1 as they have not so far been adopted to a significant extent.

12.2 Quantities of sludge

The daily production of sludge may be estimated from

$$Q\{k_1 s + (1 - k_2)Ybk_3\} \text{ kg solids/d}$$

in which Q (Ml/d) is the mean flow of sewage, s and b (mg/l) are the mean concentrations of suspended solids and BOD respectively, Y is the yield coefficient for conversion of BOD to surplus secondary sludge at the oxidation stage, k_1 and k_2 are the fractions of suspended solids and BOD removed by primary sedimentation and k_3 is the fraction of BOD removed in the oxidation stage. The factors k_1 and k_2 are roughly $\frac{2}{3}$ and $\frac{1}{3}$ respectively for conventional primary sedimentation (see p. 219), and k_3 is 0.9 to 0.95 except for partial treatment. For an activated sludge plant Y is 0.5 to 1.0 (see p. 238) and for single-pass filters 0.3 to 0.5 (see p. 271). If Y is considered to represent the total growth the suspended solids lost in the effluent (Qs_e) should be subtracted from the above expression.

The dry solids have a density of 1300 to 1500 kg/m³.

Primary sludge has a moisture content of 92% to 97% when it is withdrawn from the sedimentation tank but the fraction of total sludge withdrawn which consists of secondary sludge has a higher moisture content of 97% to 99%. The higher values are found where tanks are mechanically de-sludged while full and where sludge consolidation pockets in the tanks are minimal. Surplus activated sludges tend to have higher moisture contents than humus sludges. The moisture content of the mixture of primary and secondary sludges may be estimated by taking a weighted average of the values given.

It can be deduced that the mean flow of wet sludge to be treated is about 1% to 2% of the flow of sewage. It is not a steady flow however as de-sludging is performed at about daily or twice-daily intervals.

12.3 Sludge-drying beds

In this process a layer of wet sludge 200 to 250 mm thick is spread over a drainage layer of ashes or other granular material. Water is removed partly by evaporation and partly by drainage. When the sludge becomes dry enough to dig it is carted away for agricultural use, incineration or dumping.

Two layers of ashes are provided. The lower layer is coarse, to assist drainage, while the upper layer is fine, to retain sludge particles. Some fine ash is inevitably removed with the sludge and has to be replaced. Butt-jointed field drains under the ashes channel the drainage to manholes, and a system of impervious pipes connects these to a sump. The liquor is pumped to the works inlet for treatment.

It is necessary to provide a concrete base under the ashes and to surround the beds with brick or concrete walls so that liquor is prevented from contaminating ground water or reaching the river.

Wet sludge is distributed to the beds by a system of cast-iron pipes with valved branches to the individual beds. The branches discharge on to concrete slabs to avoid disturbance of the bed. Liquor collecting on the surface of the sludge is run off over adjustable weirs in the walls. The chambers behind the weirs may be the manholes for the under-drains.

The area required for sludge drying depends on the draining properties of the sludge and on the local climate. For undigested sludge in Britain the allowance is in the range of 0.12 to 0.37 sq m per head of population served. The lower values are applicable where the climate is dry or where the sludge is pre-treated chemically to improve drying properties by increasing particle size. The area may be reduced to as little as one-third if the sludge has been digested. Secondary (humus or surplus activated) sludges are more difficult to dry than primary sludges and it is sometimes advantageous to condition these with coagulants before mixing.

Swanwick[1] has evolved a method of estimating required drying area to suit local conditions. Samples of the sludge are tested to assess its draining properties. Monthly rainfall records and evaporation estimated from other meteorological records for the locality are then used to carry out an operational study of the drying of the sludge. From this the maximum time that the beds will be occupied by a charge of sludge can be estimated and hence the area needed.

The critical time for sludge drying is in winter. More drying area is needed in the north-west of Britain than in the south-east as rainfall is higher and evaporation is lower during the winter months. It may be advisable to use records for a difficult year rather than long-term average values.

Unless drying area is provided generously in the first instance, space should be left for convenient extensions. There may be a small advantage in siting drying beds so that they are not sheltered from the wind.

Drying beds have occasionally been protected from rain by moveable glass roofs but the extra cost is generally considered to be more than the saving from the reduction in area. Transparent plastic materials have a possible application.

The size of the units which make up the drying area should be such that each unit can be charged in a single operation. The size is therefore related to the volume of storage capacity in the primary sedimentation tanks. The total area must be divided into several units. At any given time dried sludge will be lifted from one or two units while the others are occupied with sludge at different stages of drying.

Manual sludge-lifting requires a good deal of labour, and machines which travel along the bed lifting sludge to a conveyor offer an alternative. Where a machine is to be used, the precise width of bed will be determined by the choice of equipment.

Paved access is needed to all parts of the drying area, and paved stockpiling space may have to be provided.

12.4 Anaerobic digestion of sludge

When sewage is kept in an environment without oxygen, bacterial action develops and results in some decomposition of the organic material into methane (CH_4), carbon dioxide and water, together with smaller quantities of ammonia, hydrogen sulphide and hydrogen. Commonly the organic solid content is reduced by about 50%. The odours associated with raw sludge are in part due to the occurrence of digestion within it, so sludge which has been digested deliberately is less odorous.

Anaerobic digestion occurs in two phases. In the first phase, fats, carbo-hydrates and proteins are converted to organic acids (mainly the lower fatty acids) and alcohols. The bacteria concerned are facultative, that is they can exist in either aerobic or anaerobic environments. Digestion occurs when conditions are anaerobic. Many different species of bacteria are involved. In the second step, anaerobic bacteria, the methane formers, break down the acids and alcohols into methane, carbon dioxide and water. Subsidiary bac-terial activity results in conversion of sulphur compounds and some others.

As in all bacterial fermentation processes, the population of bacteria increases during the process, using the available organic material as food to produce growth and energy. The rates of assimilation and growth are much lower for anaerobic than for aerobic processes as the energy yield of the reaction is lower. As in aerobic processes some of the cells produced may be used endogenously in the later stages.

Methane organisms function best when the pH value is about 7. Conse-quently, the two steps of digestion must be in balance with each other. If acid is being formed more quickly than it is being used, methane production will be decreased. It is sometimes necessary to correct imbalance by adding lime to adjust the pH value, though this has to be done with caution as there are other effects.

There are many chemicals which will inhibit bacterial action or prove toxic if in sufficient concentration, though in a few cases bacteria can adapt them-selves to the new conditions.

A number of organic compounds (mainly halogen compounds) which are not uncommon in industry have been shown to decrease gas production even when present in very low concentrations. Zinc, copper and nickel are toxic at low concentrations though their presence in a trade discharge may not be harmful if they are precipitated as sulphides before the biological process is reached. Some constituents of synthetic detergents can be harmful but these effects can be cured, at a price, by the addition of chemicals to the sludge.[2]

The problems arising from the increasing amounts of toxic materials in sewage make laboratory or pilot-scale work advisable before adopting sludge digestion. These compounds affect the aerobic processes of sewage treat-ment, but to a smaller degree, as the aerobic processes are rather more tolerant. The difficulties arising in sludge digestion are therefore unlikely to disappear through the exclusion of possibly toxic material from sewage.

The rate at which sludge digestion processes occur depends on tempera-ture but not in a simple fashion. There are optimal temperatures at which the processes occur most rapidly, the rate increasing as temperature increases to the optimum and then decreasing as temperature is increased beyond the optimum. Mesophilic digestion occurs between 16 °C and 38 °C with an optimum, varying with the sludge, between 30 °C and 35 °C. Thermophilic digestion, which is more rapid, occurs between 38 °C and 65 °C with an optimum at about 55 °C. In addition, cold digestion, a relatively slow process, occurs between 7 °C and 16 °C. Cold digestion requires too much capacity to be deliberately practised except perhaps at very small works where the cost of tankage is compensated for by the avoidance of the cost and complication of heating equipment.

Compared with mesophilic digestion, the provision of extra heat for the thermophilic process is normally considered to be uneconomic and there is the further disadvantage that the supernatant liquor from the thermophilic process is more difficult to treat. The mesophilic process is therefore the one which is usually adopted.

If sludge is left to digest undisturbed in a closed, heated tank, stratification inevitably occurs. In the middle there will be a layer of actively digesting sludge above which supernatant liquor will gradually accumulate. Floating on top is a scum which contains fibrous material and can become a thick mat restricting the passage of gases to the space under the roof. Stable, digested sludge settles in a layer under the digesting portion and below this, on the floor of the tank, any grit which was not removed from the raw sewage will accumulate. Grit can be removed only by digging after the sludge has been drawn off.

In principle, such a tank could be operated on the fill-and-draw system, raw sludge being added day by day until the tank was full and then left until digestion was complete. It would be necessary to 'seed' the raw sludge with actively digesting sludge to ensure that the process started readily. In practice, once the tank was full it would be possible to feed raw sludge to it daily by withdrawing digested sludge and supernatant liquor to make room for the added sludge. In this way, the tank would be kept full and would be emptied only when the necessity to remove scum and grit arose. An unmixed tank would be entirely satisfactory if it were not for the presence of scum and grit and provided that raw sludge were properly seeded by being intimately mixed with actively digesting sludge.

Tanks in which the contents are continuously mixed are commonly preferred. The contents of such tanks are not stratified but consist of a uniform mixture of sludge in varying stages of digestion, supernatant liquor, scum-forming material and, when present, grit. Seeding occurs inevitably.

If the tank has a fixed roof, the volume of material in the tank remains nearly constant and some of the contents are displaced, whenever raw sludge is pumped in. Sludge discharged from the digester has the same composition as the mixture within the tank and contains fractions of undigested and partially digested sludge. The composition of the tank contents attains an equilibrium due to the mixing. Of the sludge put into a mixed digester, some fractions remain for less than the average detention period (volume/rate of input), others for longer than the average. The distribution of retention time is similar to a negative exponential curve. The mean retention time aimed at for mesophilic digestion is about 30 days, based on the input rate of raw sludge, though a shorter detention time is sometimes adopted.

Digesters with floating roofs have variable capacity. The variable part of the capacity amounts to about one week's input of raw sludge. Sludge is pumped in daily but need only be withdrawn at weekly intervals.

The separation of supernatant liquor occurs mainly in the later stages of digestion. Sludge discharged from a mixed digester is normally put into a secondary, unheated digester where some digestion continues and liquor separates and can be decanted off. Little gas is evolved at this stage and the tanks are unroofed. They serve also as balancing storage between the heated digesters and the drying beds or other means of further drying or disposal.

In Britain primary heated digesters are designed to deal with about 1 kg of organic solids per m^3 per day, implying a capacity of about 0.056 m^3 per head of population served. Secondary, cold digesters have a capacity of 0.03 to 0.06 m^3 per head. The required capacity depends on the diet of the population. Thus in South Africa[3] it has been found that primary digester capacity should be increased to about 0.15 m^3 per head of African population.

Heated digestion tanks are circular on plan and with sloping floors and dome-shaped roofs. This approximation to a spherical shape helps to reduce the heat loss and it is not uncommon to bank earth round the digesters as insulation.

In some early digesters, pipes in which hot water circulated were attached to the inside of the walls. These had the disadvantage that dried sludge caked on to the pipes and tended to reduce the extent to which the heat could be transferred. A later development was to provide grids of heating pipes which rotated about the vertical axis of the digester. These achieved a small amount of mixing of the contents but had mechanical complications. In some cases the accumulation of grit near the bottom of the digesters ultimately prevented the rotation of the heating grids. It is now common to circulate sludge through an external sludge heater. Heat is provided by means of hot water from a boiler fired with sludge gas. Means of heating with an independent fuel are needed in addition.

Where sludge gas is used for power production, heat is recovered from the engine exhaust and from its water-cooling and oil-cooling circuits by circulating the sludge heating water through heat exchangers before boosting the temperature with the boiler. Independent engine water and oil cooling (by fan or cooling tower) has to be provided for use when sludge is not being heated. There is a great deal of complication in heating and cooling circuits in such cases and, since several different sets of extreme circumstances have to be provided for, some of the plant is used only rarely. This is a factor which contributes to the high cost of generating power from sludge gas. Comparative estimates often show that it would be cheaper to purchase electricity from a supply authority, except at very large installations.

The continuous pumping of a sludge through an external heater achieves some mixing of the contents of the digester, and the inlets and outlets are always placed so as to encourage this. In some digesters, multiple inlets and outlets are provided and circulation is relied upon to provide the whole of the mixing.

The amount of circulation required for adequate mixing may be in excess

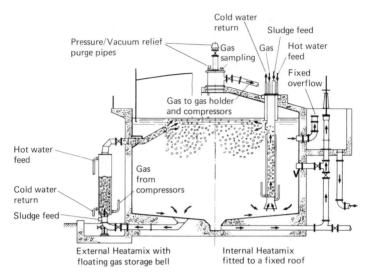

Fig. 12.2 Alternative arrangements of sludge digestion tanks employing Heatamix units (*Courtesy Simon-Hartley, Ltd*)

of that which would be required if the minimum size of external heating plant were the only criterion. Consequently, some sludge digestion tanks have provision for mixing which is quite independent of the heating arrangements, for example by re-circulating sludge gas through the digesting sludge.

If sludge gas is to be used for power production, a gas holder of about one day's capacity is needed. The rate of gas production is 0.02 to 0.03 m³ per day per head of population and its calorific value is say 21 000 to 24 000 kJ per m³.

It is feasible to provide a gas holder bell on top of the digester, the digester acting as the tank. The proportions of the bell often make this uneconomic and it may be better to provide a separate, small, guide-framed gas holder with its own water tank. Even if there are a number of digesters, a single holder is usually all that is required. A safe-burner is needed to dispose of surplus gas and will require its own well ventilated building of light cladding on a steel frame. The gasholder, surplus burner building and other buildings should all be separate from each other and should have spaces between them of not less than 7.5 m. Methane detectors may be required at certain points but ample ventilation is also an important safety measure. Gas pipes should be above ground. The boiler room must be isolated by a continuous solid wall with no openings to adjacent parts of the building. Health and Safety Executive guidelines must be followed in these and other matters relating to the gas installation.

The need for safety precautions around a sludge digestion plant is obvious but there are many other places in wastewater systems where anaerobic bio-degradation may occur leading to risks of explosive or asphyxiating concentrations of gas if ventilation is inadequate.[4, 5]

The use of sludge gas for power production became feasible with the development of dual-fuel engines which could run on oil when insufficient gas was available. A small percentage of oil is used in these engines when they are running on gas. At a works treating domestic sewage, heated sludge digestion normally produces enough gas to heat the digesters and to power an activated-sludge plant. The presence of trade waste may add to the power requirements for oxidation without adding to the gas production, which may be reduced if toxic agents are present.

It is usual for the engines to drive generators rather than to use the power directly. This enables operation to be more flexible. Alternating current is normally preferred although there are circumstances (for example, variable speed pumping) in which the decision not to use direct current is sometimes taken with reluctance. In some cases, electrical power purchased from the supply authority is used for part of the load or for standby.

Though a good deal of help is available from contractors for engines and electrical equipment at the design stage, the co-operation of a consultant must usually be sought unless the sewage-treatment department is large enough to have its own mechanical and electrical design section.

Bottled sludge gas has been used for driving road vehicles and, in wartime, for incendiary bombs. It is used in South Africa as a source of cyanide for the gold-mining industry.[3]

12.5 Conditioning and elutriation

The addition of compounds of aluminium, iron and some other metals to sludge makes the extraction of water easier. There is some controversy about the exact manner in which such compounds act but it is generally accepted that the charges on the metallic ions neutralise surface charges on the small sludge particles.

Even though the constituents of the agents used are often industrial by-products, their cost is high and it is necessary to carry out preliminary tests to find optimum doses and to select the cheapest alternative.

Crude ferrous sulphate is a steelworks by-product but needs to be chlorinated in solution. The cost of chlorination plant, and the highly corrosive nature of the final solution — known as chlorinated copperas, add to the expense of using this. Alternatives such as aluminium chlorohydrate are more expensive but are less corrosive and are preferable in some cases. Organic intermediate products of the detergent industry known as synthetic poly-electrolytes are another possibility.

The effectiveness of conditioning chemicals depends on the pH value of the sludge and it may be worthwhile to adjust pH to an optimal value before using the conditioning agent.

Particularly at plants where filter presses are used it is common to use milk of lime as a conditioner.

In the early days of sewage treatment it was considered that a precipitating agent should be added to sewage before treatment in sedimentation tanks. Lime was often used in relatively large quantities. This increased the amount of sludge to be treated. It is now considered preferable to allow sedimentation to occur without chemical aid and to apply coagulants to the resulting sludge.

The facilities needed for conditioning consist of storage space and handling equipment for the dry chemicals, mixers and solution tanks large enough to accommodate the dose for a batch of sludge, and tanks with stirrers for mixing the solution with the sludge. Chlorination equipment may be needed in addition. The sludge tanks are usually in the open, alongside a building accommodating the chemical store and dosing equipment.

An elutriator consists of an upward-flow tank in which water or, more usually, clarified sewage effluent is used to wash out the finest particles from the sludge. Removing fine particles reduces the required dose of coagulant. The elutriation of digested sludge washes away some products of digestion which would react with or reduce the effectiveness of the coagulant.

12.6 Heat treatment

If sewage sludge is kept at a temperature of 180 °C to 200 °C for a period of about half an hour, the gel-like structure is broken up and coagulation will occur fairly readily. The Porteus process employed this principle on a batch system, using steam injection as the method of heating. William E. Farrer Ltd., after becoming proprietors of this process, developed it to work continuously and eliminated the need for steam injection which gave rise to odour problems. Sludge from a storage tank is macerated, to reduce material which might cause blockages, and pumped through heating and reaction vessels in which pressure is maintained between pre-set limits by control valves.

The sludge passes firstly through a pre-heating vessel in which it is heated by sludge recirculated from the reaction vessel. At the next stage the temperature is raised to about 190 °C by means of hot water under pressure circulated from an external oil, gas or solid-fuel fired boiler. The sludge then passes through the reaction vessel, where it remains for about half an hour, after which it is returned to the pre-heater where it is cooled in contributing its heat to the incoming sludge.

After leaving the pressurised system, some supernatant liquor can be extracted immediately in a decanting tank. This is a continuous-flow sedimentation tank in which coagulated sludge settles to the floor while supernatant liquor overflows from the surface. The liquor may have a BOD of up to 4000 mg/l, and is returned to the works inlet. The coagulated sludge has a moisture content of about 87%. Further de-watering is accomplished in a filter press.

12.7 Filtration under pressure

(1) Theory

In the vacuum filter and in the filter press, sludge is held under pressure against a membrane which retains the sludge while allowing liquor to pass through fine pores. As filtration proceeds, a mat of solid particles builds up on the membrane and gradually becomes thicker. Consequently the resistance to flow increases with time.

By analogy with laminar flow through porous media and through tubes, this process may be expressed as:

$$P = \frac{dV}{d\theta} \cdot \frac{\mu}{A} \left(\frac{rcV}{A} + R_m \right) \tag{12.1}$$

in which
- P = pressure difference causing flow
- V = volume of filtrate produced in time θ so that
- $dV/d\theta$ = instantaneous rate of flow of filtrate
- μ = viscosity of filtrate
- A = area of filter
- c = concentration of solid matter in sludge
- r = 'specific resistance' of sludge
- R_m = a measure of the resistance to flow of the membrane.

The bracketed terms are analogous to kL/d^2 for a bundle of tubes of diameter d and length L with the numerical factor k combining the constant in Poiseuille's formula and the ratio of area of flow to gross area, A. The membrane resistance factor, R_m is analogous to a fixed value of kL/d^2 but the first term in the brackets has to allow for the increase in L as the mat formed by arrested particles increases in thickness. This is provided for by the group cV/A which would be in length units if c were a volume concentration, consequently the particle density is involved in r which is analogous to $(\rho d^2)^{-1}$ and provides a measure of the combined effect of size and shape of pore spaces, porosity and the influence of compressibility of the mat upon these. Decrease in particle size and hence in pore size causes r to increase.

If a strictly consistent system of units is used r is measured in m/kg or cm/g. The former are now often used. However, in earlier work from which data are still relevant, c was measured in g/ml and P was expressed in g-force/cm² even though μ was in dyn s cm⁻² (poise). These units produce a value of r measured in (cm g-force)/(dyn g-mass) but the distinction between g-force and g-mass was not made and the values were wrongly quoted as being in cm/dyn or more usually in s²/g, with g cm s⁻² substituted for dyn.

It follows that to convert r in 's²/g' to m/kg the conversion factor is 981×10^1 or approximately 10^4.

Tests made in laboratory Buchner funnels give straight-line relationships of the form

$$\frac{\theta}{V} = bV + a$$

suggested by the integration of (12.1):

$$\frac{\theta}{V} = \frac{\mu}{PA}\left(\frac{rcV}{2A} + R_{\mathrm{m}}\right)$$

Specific resistance as defined by this formulation can therefore be found by measuring the slope b of the graph of θ/V against V:

$$r = \frac{2PA^2b}{\mu c}$$

The packing of the sludge particles in the mat affects r and depends on the pressure applied. Carman proposed the relationship $r = r'P^s$ in which s is termed the coefficient of compressibility. Tests should be carried out at different pressures to estimate a value for s. Coackley, who first applied this work to sewage sludge, quotes ranges of 0.70 to 0.86 for digested sludges and 0.60 to 0.80 for activated sludges but warns that exceptions occur.[6]

Determinations of specific resistance are easily made in the laboratory and are of great value in seeking suitable methods of conditioning sludge before filtration. Elutriation and coagulation can decrease specific resistance of digested sludges from the order of 10^{14} m/kg to the order of 10^{12} or even 10^{11} m/kg. Specific resistance, which is considered to be strongly influenced by the presence of fine particles, is a useful measure of the difficulty of extracting water by any method, though it is most closely connected physically with vacuum filtration. Yields of vacuum filters and pressing times for filter presses can be estimated from specific resistance tests but it may be advisable to carry out pilot-scale work in addition. With some sludges, filter cloths tend to become blocked and with others fine material accumulates in the grooves of filter presses.

The 'capillary suction time' (CST) test[7] is now used as a quick method of assessing filtration resistance of sludges. An open ended cylinder (18 mm bore, 25 mm high) rests on a piece of thick chromatography grade filter paper. The sludge sample is poured into the cylinder. Once the paper has become wet filtrate passes rapidly into it and a mat of sludge particles forms on its surface. The absorption capacity per unit area of paper is high so that the resistance to filtrate flow is mainly that of the mat. Filtrate is drawn outwards by the capillary suction of the paper and an advancing front of filtrate can be observed. Since the paper has a grained structure the line of the front is elliptical rather than circular and for this reason rectangular papers are used, the base of the apparatus consisting of a shallow perspex tray of the same dimensions as the standard papers.

The time measured is that for the front to travel between two prescribed distances from the cylinder. Probes in contact with the paper serve (by conductivity) to register the arrival of the front at successive points and the time interval is displayed on a dial.

CST is affected by surface tension of filtrate, temperature and solids content. Corrections to a standard temperature are made by assuming that the viscosity of the filtrate varies in the same way as that of water.

There is no general correlation between CST and specific resistance but acceptably useful correlations can be established for particular sludges which enable the CST test to be used for routine control of the addition of coagulants to improve filtrability in presses and vacuum filters.

(2) Filter presses

In the filter press, wet sludge is held under pressure between filter cloths until a sufficient amount of water has been forced through the cloth for the resulting sludge cake to be handled as a solid material.

The press consists of 50 to 100 recessed cast-iron trays hanging from an overhead beam. The trays are 0.9 m to 1.2 m square and the filter cloths are trapped between them. Hydraulic rams hold the trays tightly together. Sludge is delivered under pressure to the spaces between the cloths by way of holes in the centres of the plates, sleeved with filter cloth. A pressure of about 700 kN/m^2 is required and is provided by sludge rams consisting of steel tanks in which pressure is maintained by compressed air. The trays support the filter cloths and are grooved to conduct liquor pressed from the sludge to the drainage pipes. After the sludge has been pressed for several hours, the trays are moved apart successively and the dried cake is dropped into a receiving hopper. The thickness of the cake is nominally 25 or 40 mm. Pressing reduces the moisture content to between 55% and 75%.

Pressing is a batch process and it is usual to have several presses which are charged in turn. The sequence of operations can be automatically controlled but attention is needed at the final stage of discharging cake. Pre-treatment of the sludge with milk of lime is necessary and mixing tanks are needed for this. Ferrous salts may be needed in addition to milk of lime if there is a significant quantity of activated sludge.

Filter presses are usually accommodated on the first floor of a building, together with chemical storage hoppers, coagulant tanks and filter-cloth washing equipment. The ground floor accommodates sludge rams and pipework, chemical mixing tanks and access for the removal of pressed cake.

Mixing tanks for pre-treatment are sited adjacent to the building. Raw-sludge tanks of one to two days capacity are often provided for balancing storage of sludge before pre-treatment.

(3) Vacuum filtration

The principle of operation of the rotary vacuum filter is shown in Fig. 12.3. The drum is 1.8 to 3 m in diameter and up to 3.7 m long. It is covered with cloth which forms the filtering medium. The cloth is supported by a number of panels faced with perforated metal or plastic. Pipes lead from the panels to the axis where an arrangement of valves enables each panel to be subjected to vacuum or to outward pressure according to its position. The drum rotates at between 5 and 20 revolutions per hour depending on the characteristics of the sludge.

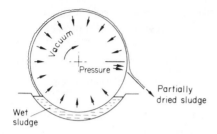

Fig. 12.3   The principle of the rotary vacuum filter

As the drum rotates, a layer of sludge is picked up from the bath and held in place by the pressure difference between the external atmosphere and the vacuum on the inside of the cloth. Under the influence of this pressure difference which is about 70 kN/m^2 water is gradually filtered through the cloth. As the panels approach the scraping blade, the vacuum is replaced by outward pressure.

The Komline–Sanderson Coilfilter (Dorr–Oliver Co. Ltd) is a rotary vacuum filter in which the filtering medium consists of stainless-steel coils in place of fabric. The coils leave the surface of the drum near the discharge level and are washed before returning to the surface.

As the filtration pressure is lower than that in filter presses, vacuum filters produce cake of higher moisture content, usually 80% to 85%.

12.8 Sludge concentrators

In the Roto-plug concentrator (Davey, Paxman & Co. Ltd), sludge is fed into a cylinder rotating on a horizontal axis. The cylinder is lined with woven fabric and liquor passes through the fabric as a plug of sludge is rolled round by the rotation of the cylinder. The sludge feed is continuous and partially dried sludge is cut away from the rolling mass by end flanges. It then falls through a hopper and is guided to a pair of rollers. One roller is of wedge wire and the other is solid. Compression between the rollers reduces the moisture content to about 70% to 80%. The product can be dried further by stacking in the open. The device has been applied to untreated primary sludge at small works. For secondary and digested sludge, fibrous matter may have to be added.

Several devices used in industry may be adaptable for extracting moisture from sewage sludge and some development work is in progress.

12.9 Composting

Putrescrible organic matter can be oxidised to an inoffensive humus of agricultural value by aerobic fermentation. For this process, air and moisture are required. Sewage sludge contains so much water that anaerobic processes are difficult to avoid. Domestic refuse is too dry for fermentation to develop effectively and may be deficient in nitrogen. The addition of sewage sludge to refuse supplies moisture and nitrogen.

The process is similar to that of the farmyard manure heap and the practice of composting sludge with refuse developed by Wylie[8] owed much to Howard's work on composting for agricultural purposes in India.

Raw refuse contains material which must be extracted before composting is undertaken. Where refuse is incinerated, it is customary to screen out fine material, to remove ferrous metals magnetically and to remove other large objects such as glass and plastics manually before burning the residue. These processes are sometimes needed if the residue is to be burnt instead of being composted, so the refuse side of a composting plant may be similar to the segregation section of a conventional refuse incinerator. Some of the material removed has scrap value.

Dust screened from the refuse is used in composting in addition to the residue of segregation. The dust possesses some compounds of agricultural value but its main value is in reducing the moisture content of mixed sludge and refuse.

Before addition of sludge, refuse has to be granulated in some form of mill to increase the surface area accessible to organic action. Green vegetable matter is readily decomposed, but bones and cellulose materials, wood and paper, require longer periods of fermentation.

The moisture content of mixed sludge and refuse needs to be in the range 40% to 60%. If the plant is to deal with the sludge and refuse from the same population, partial de-watering of the sludge mechanically, or by digestion may be needed. Alternatively, additional vegetable matter may have to be provided, or some of the sludge may have to be dealt with separately. The seasonal variation in the quantity and nature of the refuse may need to be considered.

Composting plants of widely differing forms have been successfully operated. In plants of the form adopted at Kirkconnel, Dumfriesshire, open cells 4.3 m × 4 m × 1.8 m deep are filled in 0.3 m layers with a mixture of wet sludge, pulverised residual refuse, straw and refuse dust. The wastes are mixed in chambers, with provision for drainage of surplus liquor, before being placed in the cells.

The walls and floors of the cells have ventilation holes. Fermentation in the cells occupies about 5 weeks. Temperature rises to about 80 °C during the first two days and then falls irregularly to about 30 °C. Compost is moved to maturing bays for a further period of 5 weeks. An overhead grab is used for transferring material from one stage to the next and for turning the compost.

In another form of plant the cells are one above the other in a tower. The mixed waste begins at the top of the tower and falls successively through hinged flaps to the lower floors at intervals of about a day. Fermentation in the tower occupies about a week and is followed by several weeks of maturing under cover.

The Dano process employs a long drum rotating on a horizontal axis at a rate of 20 revolutions per hour. The drum is operated fairly full of water and the mixture follows a helical path and is discharged at the end after 3 to 5 days' treatment. Sludge (or water if the stabiliser is used for refuse alone) is injected at the inlet end and air is supplied under pressure through nozzles along the length. Temperatures of about 55 °C are maintained in the middle section of the drum by aerobic activity which may be controlled by the rates of air and moisture injection. The rotation of the drum, as well as enabling material to move readily along, promotes abrasion of the solid wastes and ensures mixing of the fermenting material.

12.10 Incineration

The organic constituents of sludge are combustible provided that they are not associated with too much water. Incineration is therefore a final disposal process used after one or more water-extraction stages, as an alternative to agricultural use or dumping. It is not used widely at present, being regarded as a last resort, but improvements in methods may in the future lead to its becoming more competitive.

The simplest form of sludge incinerator is the multiple-hearth furnace. The hearths are floors in a cylindrical tower. Milled sludge cake from filter presses or other drying units is introduced at the top of the chamber and is moved over the floors by rakes, gradually falling to successively lower levels. The furnace is fired by gas or oil burners.

The Dorr–Oliver Fluo Solids system employs a fluidised bed furnace. This consists of a cylindrical vessel with its axis vertical containing a bed of sand. When the furnace is in operation, the sand is kept in a highly turbulent state of suspension by an upward current of air injected under pressure from beneath. Sludge is injected continuously to the fluidised bed and little additional fuel is needed to maintain the operating temperature of about 800 °C, as the sand retains heat. Ash resulting from the combustion of the sludge is carried upwards by the flow of hot gases and is exhausted near the top of the furnace. The exhaust gases are used to preheat the air supply and then pass to a scrubber in which the products of combustion are transferred to a flow of treated effluent. Ash is removed from the effluent by a cyclone and the effluent is returned to the works inlet for treatment. Before incineration, grit-free, thickened sludge is treated in a centrifuge or on a vacuum filter to reduce the moisture content.

The Zimmerman, wet-oxidation, process uses wetter sludge and a lower temperature. Sludge and air under high pressure are forced through a tubular chamber at 250 °C in which oxidation occurs. After being used to pre-heat sludge, the hot gases carrying the ash are condensed in a cyclone. The condensed liquor is returned to the works inlet after the ash has been removed by settlement.

Incineration, wet-oxidation and heat-treatment coagulating processes can almost be said to form a continuous spectrum, as wet-oxidation operated at low pressure results in only partial oxidation but produces material from which water can easily be extracted.[9]

12.11 Sludge pumping

Though sludge, even at low solids content, is considerably more viscous than water, the head loss in pipelines is not very much greater and can be estimated by applying a factor to the head loss, or hydraulic gradient, calculated for water. Factors of 1.05 to 1.35 have been found but it is prudent to use 1.5 for design and even higher values have been used.

Since sludges have such high viscosities, the minimum velocity for the flow to be in the turbulent range is much higher than for water. In pipes of 157 mm diameter minimum turbulent flow velocities of 0.6 m/s have been found for sludges with 2% to 3% solids ranging to more than 2.5 m/s for sludges with 5% to 9% solids. For a particular sludge the minimum velocity varies with solids content but there are large differences between sludges from different sources.[10]

In the laminar flow range sludge flows as a plastic rather than a fluid, that is a yield value of shear stress has to be exceeded before flow will occur. For a Bingham plastic, Fig. 12.4, there is a linear relationship between shear stress and rate of shear strain, as for a Newtonian fluid, once the yield stress is overcome. Sludges are similar to Bingham plastics except that shear stress is lower at low rates of shear strain.

It is best to use the turbulent range in designing pumping systems since apart from the lower friction factors high velocities are needed to ensure that grit, grease and stringy materials are moved along. Long radius bends and radial tees should be used in the pipework. Convenient access points should be provided.

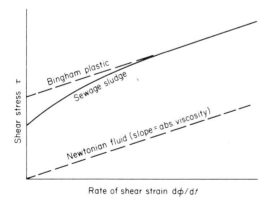

Fig. 12.4 Relationship of shear stress to rate of shear strain for sludge

Though centrifugal pumps can be used, reciprocating pumps are often preferred with air vessels to smooth delivery. Shear pins are provided to protect the machinery from the effect of stubborn blockages.

Sludge has been found to have up to 100 times the compressibility of water due to the presence of gas bubbles so there is less hazard from pressure surges than when pumping water.

Screws lifts (Section 5.7) are now the common means of recycling activated sludge.

References

1 Ministry of Technology, *Water Pollution Research*, **1965**, 89–93, H.M.S.O., 1966.
2 Downing, A.L. and Swanwick, J.D., Treatment and disposal of sewage sludge, *J. Inst. Mun. Engrs*, **94**, 81–86, March 1967.
3 Hamlin, E.J., Sewerage and sewage disposal in sub-tropical countries with special reference to South Africa and Mauritius, *Proc. Inst. Civ. Engrs*, Pt. 1, **1**, 561–603, Sept. 1952.
4 Woods, D.R., Potential hazards to operatives in the water industry, *Proc. Inst. Civ. Engrs*, **80**, 53–67, Pt. 1, Feb. 1986.
5 British Standards Institution, BS 5345. Code of Practice for the selection, installation and maintenance of electrical apparatus for use in potentially hazardous atmospheres, B.S.I. London.
6 Coackley, P., Principles of vacuum filtration and their application to sludge drying problems, in *Waste Treatment*, P.C.G. Isaac, (Ed.), 317–334, Pergamon Press, Oxford, 1960.
7 Baskerville, R.C. and Gale, R.S., A simple instrument for determining the filtrability of sewage sludges, *Wat. Pollut. Control*, **67**, 233–241, No 2, 1968.
8 Wylie, J.C., *Fertility from Town Wastes*, Faber, 1955.
9 Camp, I.C., Examples of sewage sludge incineration in the U.K., *Proc. Inst. Civ. Engrs*, **56**, 49–62, Pt. 1, February 1974.
10 Hayes, J., Flaxman, E.W. and Scivier, J.B. A comprehensive scheme for sewage sludge disposal in North West England, *Proc. Inst. Civ. Engrs*, **55**, 1–21, Pt. 2, March 1973.

13

Tertiary treatment

13.1 Introduction

Conventional works employing biological filters or activated sludge are capable of producing effluents of 20 mg/l BOD and 30 mg/l SS (for 95% of the time). If the receiving water affords little dilution or there is a need to preserve a particularly high quality, an attempt must be made to obtain effluent concentrations below the 20/30 standard. The processes used are described as tertiary treatment.[1]

Some of the processes — microstraining and the various forms of rapid sand filtration — are used in the treatment of water supplies and their main objective is the removal of suspended matter resulting in an incidental removal of some BOD. They provide no substitute for efficient bio-oxidation and indeed it is usually considered that full nitrification should be achieved at the bio-oxidation stage wherever an effluent demanding tertiary treatment is required. Nevertheless the use of microstrainers to remove humus from filter effluent is technically feasible.[2]

A fourth process — slow sand filtration — is also borrowed from water treatment and enables some biological oxidation to occur but it is costly in space and in labour so it can hardly be considered as a means of reducing the load on the main bio-oxidation process.

Pebble-bed clarifiers are a means of improving the effluent from the humus tanks in which they are customarily installed and are again mainly a suspended solids removal expedient.

Allowing effluent to flow over grass plots is a simple way of achieving improvement. Though in the present context it is considered as a process applied to settled effluent, on a domestic scale it can be used in place of a humus tank thereby avoiding the need to remove and dispose of humus sludge.

Lagoon treatment of final effluent has several biological implications in common with algal ponds (see p. 274) used for full treatment.

Each of the processes is described below. For those processes noted above as being used in water treatment further detail will be found in books on water supply.[3]

13.2 Microstrainers

The filter element consists of woven wirecloth with apertures of 15 to 65 micron according to the mesh used. Even smaller particles are arrested due to the build up of a mat of solids on the cloth. The strainer is in the form of a drum with its axis horizontal. The upper part of the drum is above water level. Flow through the wirecloth forming the drum is outwards, the channel feeding the strainer being in line with the drum axis. The chamber in which the drum is installed receives the filtered liquid which then passes into the effluent channel. When the head difference across the wirecloth exceeds about 150 mm a weir is overtopped and excess flow is bypassed.

The drum rotates continuously at a peripheral speed up to 0.5 m/s. Solid matter is removed from the cloth by means of wash water (effluent) jets above the top of the drum. A trough inside the drum, beneath the jets, receives the wash water which is returned to the head of the works and amounts to about 2% of the effluent. Ultra-violet lamps beside the water jets prevent biological growth on the wirecloth.

A laboratory testing technique is available for determining the rate of application of flow to unit area of fabric and the speed of rotation. For tertiary treatment the rate of maximum flow is typically in the range 300 to 700 m^3/d per m^2 of fabric giving a reduction in SS of $\frac{2}{3}$ to $\frac{3}{4}$ accompanied by a reduction in BOD of $\frac{1}{4}$ to $\frac{1}{2}$. If microstrainers are to be used in substitution for humus tank capacity rates as low as 200 m^3/m^{2}d are likely to be needed. Site area for an installation is about 90% of the area of fabric installed.

13.3 Rapid gravity sand filters

Here the filtering element consists of a bed of 0.8 to 1.7 mm sand 0.6 to 1.2 m deep operated in a submerged condition with water level about 0.5 m above the top of the sand. Flow is downwards. The sand bed is contained in a rectangular concrete tank in the floor of which are nozzles connected to pipes below the surface of the floor. A layer of graded gravel on the floor supports the sand, enables the flow to be more uniform and prevents the loss of sand. Flow enters the filter through a penstock above water level.

The action of a sand filter is not entirely a sieving process. The interstices of the bed act to some extent as tiny settling tanks so that particles smaller than the spaces between the sand grains are removed from the flow. In addition accumulation of material in the surface layer reduces the size of the pore spaces.

Accumulation of material increases the head loss through the sand bed and when head loss reaches $2\frac{1}{2}$ to 3 m the flow becomes impractically low. The filter is then taken out of service and cleaned. To do this the water level is drained down and then wash water is forced through the floor nozzles at a rate of about 800 m^3/m^{2}d. This is sufficient to expand the sand bed by suspension of the grains. The water level rises higher than the normal filtering level and trough weirs across the upper part of the filter tank remove the wash water. Washing is needed about once a day and uses about $2\frac{1}{2}\%$ of the effluent which is returned to the head of the works after use. Because of the necessity for washing, filters are installed in groups so that they can be taken out of service in turn. In most cases it is considered advisable to use compressed air to scour the bed at the beginning of the washing cycle.

In order to ensure that flow passes through filters at a constant rate during treatment, control valves are provided at the outlet. In effect these impose a head loss on a clean filter which is automatically reduced as the head loss through the bed increases.

In principle a rapid gravity filter installation needs frequent attendance but labour is reduced by using motorised valves so that the washing operations may be carried out from a console. The full cycle can be automated to reduce attendance even further.

There is an inevitable head loss through a rapid gravity filter installation of 3 m or more so the whole flow may have to be pumped on a flat site.

Treatment rates at maximum flow vary from 100 to 250 m^3/m^2d, the lower rates achieving SS removal as high as 90% with 70% removal of BOD.

Pressure filters are similar in principle to rapid gravity filters but the bed is contained in a steel plate cylinder with its axis vertical (small units) or horizontal (large units) instead of in an open concrete tank.

13.4 Upward-flow ('immedium') sand filters

The washing process described above will stratify a sand bed into layers of different grain size. The smaller particles having lower settling velocities relative to the fluid (see p. 187) are left at the top of the bed, particle size increasing with depth. For this reason the sand used for rapid gravity (downward) flow filters is kept within a relatively narrow size range otherwise the interstices at the top of the bed would be too small. However, if a filter is used with the flow receiving treatment passing upwards gradation of grain size is beneficial as the interstices become progressively smaller in the direction of flow. The whole depth of the medium is then effective whereas in a downward flow filter removal of suspended matter is mainly confined to the surface layer of the bed.

Upward flow enables rates of flow per unit area to be doubled as compared with downward flow rapid gravity or pressure filters. A loading of 400 m^3/m^2d is typical for well nitrified effluent.

13.5 Slow sand filters

Like rapid gravity filters, slow filters consist of a submerged sand bed through which water flows downwards but controlled at a much slower rate of the order of 2 to 3 m^3/m^2d at maximum flow. At this rate of flow washing by upward flow at frequent intervals is unnecessary. Most of the suspended matter is accumulated in the top few centimetres of the bed and it is sufficient to remove this depth of sand, after lowering the water level, at intervals of a few weeks. The sand removed is replaced with washed sand but the initial thickness of the bed is made large enough for replacement to be needed less frequently than removal. The depth of the filter is usually about half that of a rapid filter so there is less overall head loss.

Since a slow filter is disturbed infrequently there is an opportunity for micro-organisms to grow. When used for water supply treatment some degree of oxidation occurs but the improvement in BOD due to oxidation when slow filters are used as tertiary treatment is not readily distinguished from that due to suspended solids removal and may not be very large. Slow filters remove about 60% of SS and 40% of BOD from sewage effluent.

Owing to the low rate of flow, slow filters occupy a large area — about ten times that of primary sedimentation tanks — and even though the form of construction is simple they are expensive. They are unlikely to be a proposition except on small works.

13.6 Pebble-bed clarifiers

These were devised by D.H. Banks. A bed of 6 to 25 mm gravel, 0.1 to 0.3 m thick is supported on wire mesh just below water level in the humus tanks so that flow has to move up through the bed to reach the weir. They have been applied to upward flow[4] as well as to rectangular humus tanks. Some experimental results are available linking flow rate, size of medium and depth of bed to suspended solids removal.

Typical rates of treatment at maximum flow range from 14 to 40 m^3/m^2d giving SS and BOD from 50 to 60% and 25 to 40% respectively.

The pebble bed is easily washed by lowering the water level in the tank below the bed and allowing the effluent to run out carrying the solids with it, assisted if necessary by a jet of water or effluent. The solid matter thus mixes with the humus and does not have to be dealt with separately. When solids have been left in the bed for some time, denitrification has been observed to occur and this may be capable of beneficial exploitation.

13.7 Settlement

The Aldwarke Works at Rotherham[6] had a group of four storm tanks which showed signs of structural damage caused by flotation pressures. In order to ameliorate this problem three of the tanks were kept full by pumping effluent through them. The fourth was kept empty to receive the first flush of storm water. Its outlet weir was raised and automatic controls were provided so that when the water level reached the original top water level the effluent pumps were stopped and storm water began to flow into the three tanks displacing the effluent remaining in them. The main flow of effluent passed directly to the river during the period when storm water was passing through the three tanks. After a storm all four tanks were desludged and then emptied by pumping their contents to the head of the works. Use of the three tanks for effluent settlement was then resumed.

During the evaluation period effluent SS removals of 35.7% to 43% were experienced accompanied by BOD removals of 25% though in one period these removals were 26.7% and 12.5% respectively. Since the normal storage capacity of the storm tanks was not available the frequency and duration of storm water discharge to the river via the tanks increased.

If the storm tanks at an entirely new works were to be used in this way it would seem to be best to site them so that effluent flowed to them by gravity. The saving in effluent pumping costs would more than cover the cost of the increased lift when emptying the tanks after storms.

Simple sedimentation as a means of tertiary treatment has been employed at some small works in Shropshire[7] where the effluent channels were generously proportioned to give low velocity and long detention periods so as to become long sedimentation tanks. Weirs were used to divide the channels into sections.

13.8 Grass plots

Effluent is fed by a system of channels to a gently sloping area of grassland and is collected by channels at the foot of the slope. If the slope is too steep the flow becomes localised. The maximum slope recommended is 1 in 60. The area may be specially prepared and seeded but natural grassland has been found to be satisfactory. The grass does not need to be kept short and mowing is necessary only to prevent the larger weeds from becoming established. The total area is divided into sections which are taken out of service in turn for mowing after drying out. Some accumulated dry solids have to be removed. Grass cuttings must also be removed.[8]

Rates of treatment of mean daily flow have been quoted ranging from 0.12 to 0.8 m^3/m^2d. Removals of SS range from 60% to 75%. About 55% of BOD is removed.

13.9 Lagoons

These consist of artificial unlined lakes about 1 metre deep. It is usually considered best to have several lagoons in series so that short-circuiting is prevented and the full capacity can be effective in balancing fluctuations in effluent quality.

If the mean detention time is up to about $2\frac{1}{4}$ days the lagoon acts mainly as a settling basin removing 25% to 70% of suspended solids and more than 30% of BOD. The surface loading rate corresponding to $2\frac{1}{4}$ day detention is

0.44 m³/m²d for a 1 m deep lagoon. Inevitably sludge accumulates in the lagoon but appears to reach an equilibrium possibly through biological activity so that de-sludging is not likely to be needed. Occasionally lagoons have been emptied and the sludge dealt with by ploughing it into the earth bed of the lagoon after it has become dry enough to bear a tractor.

With detention times in excess of $2\frac{1}{2}$ days algal growth occurs in summer. The algae in the effluent result in suspended solids concentrations being higher than in the inflow. However, there are beneficial effects of algal growth in that nitrogen and phosphorous are used by algae during the growing season and are thus partly removed from the effluent. At least 10 days retention is needed for maximum algal growth. There is little removal of phosphorous in winter in the U.K.[9]

Lagoons large enough to enable algae to become established have much in common with stabilisation ponds (Section 11.20).

Lagoon outlets need scum baffles as trouble with rising sludge is sometimes experienced.

Fish frequently thrive in lagoons which also attract bird life so that large lagoon systems have potential as nature reserves. Seasonally, insects may be a nuisance.

References

1 Institute of Water Pollution Control, *Unit processes: tertiary treatment and advanced waste water treatment*, 1974.
2 Cassidy, J.E., The operation of the Hazlewood Lane Works of the Bracknell Development Corporation, *J. Proc. Inst. Sew. Purif.*, **3**, 276–293, 1960.
3 Twort, A.C., Law, F.M. and Crowley, F.W., *Water Supply*, 3rd Ed., Edward Arnold, London, 1985.
4 Banks, D.H., Small sewage works that function satisfactorily, *Wat. Pollut. Control*, **75**, No. 2, 162–175, 1976.
5 Pullen, K.G., Experiences with tertiary treatment at sewage works of the Lichfield RDC, *Wat. Pollut. Control*, **72**, Pt. 1, 52–59, 1973.
6 O'Neill, J., The use of storm tanks for the tertiary treatment of sewage, *Wat. Pollut. Control*, **72**, No. 1, 87–90, 1973.
7 Ward, P., Experiences with tertiary treatment in Shropshire, *Wat. Pollut. Control*, **72**, 60–70, Pt. 1, 1973.
8 Hopper, H.H., Operating experiences with tertiary treatment at the sewage works of Cannock Urban District Council, *Wat. Pollut. Control*, **72**, 46–51, Pt. 1, 1973.
9 Fish, H., Some investigations of tertiary methods of treatment, *J. Inst. Pub. Hlth Engrs*, **65**, 33–47, 1966.

14

Overall design of treatment works

14.1 Stages of design

In common with other civil engineering projects, design work on sewage-treatment schemes is normally carried out in three stages. Each stage results in a complete set of drawings, estimates and other documents, but the degree of detail increases as subsequent stages are reached.

(1) Preliminary design

The object is to show that a scheme is feasible and to estimate its probable cost so that the authority can decide whether to proceed further. The drawings, estimates and report of the preliminary design will be used as the basis of the authority's programme of expenditure.

It is at the preliminary design stage that most of the data (Section 14.2) will be collected, though very detailed investigations will be left to the next stage. Often several alternative sites will have to be compared and fairly firm decisions reached as to the treatment processes; for example, whether to use filters or activated sludge, whether sedimentation tanks are to be circular or rectangular, what method of sludge treatment and disposal is to be adopted. Conceivably, the best combinations may be different at the alternative sites.

For the site and scheme finally selected, outline drawings showing the layout and the leading dimensions of the units will be prepared. Since the relative levels of the units will be needed before excavation quantities can be estimated, hydraulic considerations enter but experience and comparison with other schemes will avoid the need for elaborate calculations.

The main structural problems will have to be settled and the thicknesses of walls, floors, etc. estimated. Here again, experience and comparison may be sufficient, but some calculations will be needed. As most of the structures are to retain water, thickness of elements, which depends mainly on bending moments, will be determined before calculating steel requirements and the calculations can stop when the thicknesses have been roughly assessed. Steel requirements can be estimated by applying factors to the volumes of concrete.

Quotations will be needed for machinery and other manufactured equipment but these will be obtained merely to have a realistic estimate, and the items whose prices have been used will not necessarily be the ones which are finally installed. The total estimated cost will be based on gross rather than detailed quantities and the prices should therefore be adequate to cover minor features which have not been specifically measured. A Water Research Centre publication provides formulae based on analysis of contracts, enabling rapid estimates to be made.

Purely from the engineering point of view it is best to use current prices but the authority will wish to know what the scheme is actually going to cost, not what it would cost if it had been constructed immediately. It is rare for a scheme to be constructed within 5 years of the first estimate and the period is often much longer. A sum should therefore be quite explicitly added to cover increases in costs with time. The total of the current cost and of the possible increase is the figure which the authority should consider.

An underlying principle in preparing the preliminary design and estimate is that these should cover a scheme which is at least as elaborate and costly as that which will be finally built. If this is done, it is possible that more detailed work at a later stage will reveal some possible economies. The preliminary stage is not the time for seeking ingenious ways of reducing the cost. They may turn out to be less feasible at a later stage when the requirements have been more fully determined.

(2) Contract design

When the preliminary design has been fully approved, the next stage is to prepare drawings, specifications and bills of quantities for tendering purposes. Ideally these should be in sufficient detail to enable the works to be constructed without further instructions, though later modification of some details is nearly always necessary and some of the detailed design may have to be left until after the contract is let.

The first part of the work will amount to repeating the preliminary design with more exact data. More information will be available on trends in population, on flows and on trade discharges. Results will be available from laboratory investigations and pilot-scale work if these have been undertaken. More detailed site investigations and surveys will have been made.

Approval of the preliminary design is often given subject to several modifications being made and it is not uncommon for these to involve quite large changes in the design, particularly in layout.

Detailed hydraulic calculations will have to be made and precise dimensions worked out. All pumping systems and pipelines will have to be fully detailed.

Tenders will be sought for machinery and the successful tenderers nominated as subcontractors in the main contract. Final structural design and working drawings cannot be prepared until machinery has been selected and details of loads and fixing arrangements obtained, so selection of machinery has to be done at an early stage.

Though the final detailing of concrete reinforcement is not absolutely essential at the tendering stage, the structural calculations will have to be fully complete. If the delivery period for reinforcement is long, the steel may have to be detailed before the tendering stage and ordered by the authority. This may apply also to certain valves, machinery and even pipework.

Very large schemes are sometimes carried out in stages. An initial contract may be let for earthworks and site drainage, subsequent contracts being let for sedimentation tanks, activated-sludge tanks, sludge-treatment works, etc. in turn. In this case detailed design for the later contracts will proceed while constructional work is being carried out under the early ones. This spreads the design work over a longer period and enables a smaller design staff to be used. Alterations to work already constructed are almost inevitable and construction costs may be rather greater than for a single large contract. If dividing the work into separate contracts leads to several contractors working in turn, the total overheads charged under the several contracts will be greater. Often, though, the same contractor will be employed on several contracts, as the contractor who gains the first contract will have an advantage over his competitors in that his overheads for subsequent contracts can be less and in this case the cost may be little more than if a single contract had been let for the whole scheme.

There is a serious economic objection to spreading construction over a long period if works constructed and paid for in the early stages stand idle until the whole scheme is complete. This is in contrast to projects such as motorways where separate sections are of value as soon as they are complete. However, the rate at which capital can be spent and hence the total time of construction depends on wider considerations of the national economy and may be outside the authority's influence. Apart from the wider considerations there is no doubt that, for minimum cost, the whole scheme should be designed at once and then constructed in the shortest time consistent with economy.

(3) Modification during construction

Ideally, this stage should not exist. The contractor should be given full instructions in the form of drawings and specifications at the outset. However, the need for re-design and variations arises almost inevitably. Ground conditions may be different from those expected, mistakes in the original design come to light, possible improvements are noticed. Where fine detail has been left to be settled during the contract it sometimes influences detail already designed. Work arising through variations nearly always increases the overall cost, though a contingency item is provided to cover it so that the actual contract sum may not be exceeded. It is only by thoroughness at stage (2) that variations can be minimised.

An important final task of the engineer is to provide 'as built' drawings of the works, together with diagrams of circuits and valves, operating notes and details (such as drawings, parts lists and instructions) of all manufactured items installed.

14.2 Design data

(1) Area to be served

Since the necessity for a sewage-treatment scheme arises through building development, the area to be served will be defined to some extent at the outset. In spite of this, it is necessary to see whether additional land could be served by the site envisaged and whether such land is likely to be developed even if it is outside the drainage area. In contrast, a part of the area originally defined may turn out to be better served by another works. The treatment scheme finally adopted should be drawn up so that the site can ultimately be used to serve the whole of the area which can be conveniently drained to it.

(2) Population

Data on present and future population are available from town-planning departments. While these data provide reliable short-term forecasts, it should be realised that town planning is necessarily an evolutionary process, proposals are changed from time to time and some changes are beyond planning control. Consequently, the sewage-works designer often needs to do some additional forecasting, especially when deciding how much extra site area should be left for possible future extensions.

In undeveloped countries, population forecasts may have to be specially undertaken. Increase in the size of towns and cities is a world-wide phenomenon. Such increases often follow a 'compound interest' law, the rate of increase at any give time approximating to a constant percentage of the population at that time. If a rough forecast is to be made by fitting a line by eye to a graph of past records it is best to use semi-logarithmic graph paper as the fitted line can then be straight. More objective forecasts can be made by using the least-squares method to find a line of best-fit to the records. Standard errors, measuring the scatter of individual values about the line, and standard errors of estimate of the line itself can easily be found without much additional calculation and should always be considered. Several alternative forms of relationship between population and time can be tried and compared. If only 10 to 15 years of record are available the calculations are not too extensive to be done on the desk but for long records, it is worthwhile to use a computer, and several different groups of years can then be used to obtain results for comparison.

(3) Rate of flow

Where a part or the whole of the area is already sewered, data on rates of flow can be obtained by gauging. It is important to know the minimum flow as well as the mean dry-weather flow and the maximum. If population changes are expected, rates of flow on a per head per day basis will be needed.

As mentioned previously, the consumption of water closely approximates to the dry-weather flow of sewage, so it is common to obtain records of water supply. The variations in foul-sewage flow are masked by the effects of storm water in sewage-flow records so water-consumption records are often useful even where sewage flow has been recorded.

In order to estimate future rates of flow, trends in recorded flows need to be examined graphically or by curve fitting. Rates per head of population are the most useful but bulk flows are often studied in addition, especially where a separate consideration of population trends is not warranted.

If the sewage flow has been gauged, trade discharges will in a sense have been covered but each discharge and possible future discharge will have to be considered separately to see if any major changes are likely to influence the flow. Maximum rates, including those arising in emergencies, are particularly important.

(4) Strength of sewage

If the area to be served is not yet developed or is unsewered, the strength of the sewage can be assessed only by comparison with similar places. This is reliable enough for the design of works to treat domestic sewage. Each case of trade discharge has to be considered separately.

Where the area is already sewered, samples may be taken for analysis. Sampling (p. 168) needs to be thorough, so that strengths can be obtained from various states of flow, rainfall conditions, chance variations, times of

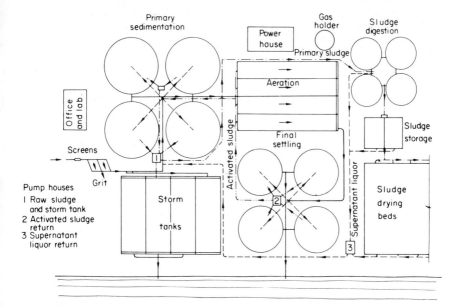

Fig. 14.1 Schematic layout of a sewage-treatment works

day and of the year. Too often, sampling is restricted to a few spot or composite samples. These are then subjected to painstaking analysis and the final results regarded as firm figures. Their true value may be dubious and indeed figures based on inadequate sampling can be quite misleading.

The object of analysis is to decide upon a suitable method of treatment and to provide data for process design. In many cases the measurement of the more important properties, for example BOD, suspended-solids and ammoniacal nitrogen concentration will in the end turn out to be the only ones which are made use of, but there is always the chance that there is something unusual about the sewage which ought to be taken into account. Consequently a very wide range of tests should be carried out on at least some of the samples. The results of these will enable an experienced sewage-works chemist to detect unusual factors. The chemist should advise on sampling and analytical procedures as well as on the interpretation of results.

In cases where abnormal features are suspected, laboratory or pilot-scale tests to discover the effects of treatment processes on the sewage may be needed, as even with wide experience it is not always possible to predict performance from analytical results. Pilot-scale work to provide data for process improvement and for the economic design of extensions is often a long-term activity at large works.

The possibility of future changes in quality must be considered. It is not unlikely that flow will increase to a greater degree than impurities so that the balance between needed hydraulic capacity and impurity capacity may gradually change. Some stages of treatment may not need extension if discharge increases without an increase in impurity load, provided that the hydraulic capacity of channels, pipes, weirs, etc. is adequate in the first place.

(5) Site conditions

In carrying out a topographical survey of the site it is worthwhile to establish several concreted reference points and temporary bench-marks at the earliest possible stage. These will then be available when trial borings are being carried out, and can be used ultimately for setting out the works. They should be clear of the area to be used for construction and it pays in the long run to have more than the minimum number of points from which measurements and levels can be taken.

Trial boreholes are often carried out under contract by specialist firms who will undertake complete site investigations, including the testing and interpretation of samples, and the preparation of a report. Alternatively, a consultant may be appointed, the contractor carrying out drilling and sampling to his instructions.

There are some features specific to sewage-treatment works which affect site investigation. Nearly all sites are close to a river and have alluvial deposits. There may be a drift-filled pre-glacial valley complicating the configuration of the rock-head. The natural ground-water level will usually be high. Most of the structures to be constructed are tanks below ground level.

Information is required as to the depth and nature of solid rock in case rock excavation is needed. Where tanks are to be founded above rock level, it is necessary to know whether piles will be needed. The loads imposed by tanks are often less than the weight of material removed to accommodate them, but piling will be necessary to prevent fracture if the tanks cannot be founded on a uniformly good stratum. If ground-water levels are high, consideration may have to be given to providing a system of under-drainage to assist construction and to reduce flotation forces. The height of the water-table above the tank floor will influence the external pressure to be used in design. Screw piles may be needed to resist flotation. A few boreholes should be lined so that they can be used to record seasonal variations in water level after the original site investigation has been carried out.

On a large scheme it is worthwhile to make two site investigations. The first series of boreholes would survey the site generally and would guide the design of a preliminary layout. More detailed investigation would then be made of the areas chosen for the main units.

Apart from topographical and geological surveys information will be needed on maximum and minimum river levels, and on the feasibility and probable cost of providing connections to the public electricity, water and telephone services.

In overseas countries, data on climate will have to be obtained.

14.3 Site layout

The main effect of layout on capital cost is in the costs of excavation, filling and the associated work of drainage and forming foundations. The effect of layout on the costs of purely structural features is small, the sizes and shapes of the units being determined by the flow to be dealt with and the processes chosen. Layout may affect costs of pipe-runs, etc. but these form only a small part of the total.

It is obvious that the stages of treatment should follow each other in a downhill direction wherever possible as each stage will need to have a lower 'top water level' than its predecessor. In any given case, there will be an optimum slope for the site which will enable the stages to be sited close to each other. Naturally this slope is flatter for a large works than for a small one. If the slope of the site is less than would be ideal, excavation can be saved by separating the stages at the expense of longer channels and pipework connecting the stages. On steep sites, the choice lies between using the high end of the site, involving deep excavations for the early stages and underbuilding for the later ones, or using a lower part of the site with less excavation and more underbuilding. The relationship between the levels of the outfall sewer and of the river will limit the choice.

Filter plants are not well suited to sites with a small overall fall, though for economy a reasonably flat area is needed for the filters themselves. If each stage is sited on ground at a suitable level, it is not uncommon for stages to be quite widely separated in a works employing filtration.

The choice of levels and positions for the main stages of treatment cannot be separated from consideration of sludge treatment. Sludge has to be moved from sedimentation tanks for treatment and the liquor removed in the de-watering process has to be delivered for treatment at a level above that of the sedimentation tanks. Pumping at some stage is unavoidable. At small works it is sometimes possible to site drying-beds at a low enough level to be charged by gravity but liquor draining from the beds has then to be pumped up to the works inlet. In other cases, sludge may be pumped to drying beds high enough for the liquor to gravitate back to the works.

Since the volume of liquor is less than that of the sludge, there is some saving in running costs if the sludge beds are at a lower rather than a higher level. On larger works, pumping sludge between various stages of treatment is often unavoidable. Whether the works is large or small, drained liquor is often gravitated to a general-purpose pumping station arranged to deal with liquor from tanks when emptied for maintenance or from storm tanks.

On a small works it is sometimes possible to arrange for a single pumping station to perform all pumping duties though care should be taken to see that anxiety to keep capital costs of pumping plant down does not lead to pumping at higher lifts than are really needed, nor to the main process units being built in uneconomic places. Some pumping duties can be adequately performed with a portable set.

There are advantages in a compact layout which may outweight the additional capital cost in excavation needed to achieve it. Such a layout saves operational labour and saves on channels and pipework. Most works require extending at some time and eventually space will be needed for reconstruction and improvement. If extension is very likely, it is best to produce an outline of the ultimate layout before detailing the more immediate requirements. This will result in space being left for expansion at the sides of the initial works while keeping the layout compact longitudinally. Space for the extension of sludge-drying areas should always be provided unless a generous area is to be allowed in the first instance. In some cases it may be best to consider providing for an eventual increase in capacity by leaving space for a more or less independent plant on another part of the site.

Wherever possible the area left undeveloped should be of such a shape that it can be used for allotments, grazing or hay, and can be fenced off from the operational part of the site. The maintenance of the 'outfield' of a sewage-works site can be a significant part of the cost of operation. The maintenance of lawns, shrubs, etc. within the operational area is worthwhile but it is not justifiable to extend these over the whole site.

Convenient road access must be provided to the sludge-drying facilities and to all points where machinery is installed. Footpath access will be needed elsewhere with ramps rather than steps wherever possible. Safety rails are essential wherever there is close access to tanks and large channels. It must be remembered that parties of visitors are shown round some sewage works quite frequently.

From the point of view of layout, every scheme is an individual problem. If an economical and satisfactory compromise is to be achieved, a good deal of thought and trial is needed. Though the final scheme will, if it is a good one, look simple and inevitable it will often have evolved from earlier ideas which would have led to untidy and uneconomic designs.

14.4 Hydraulic design

(1) Flows

The first task in designing the pipes and channels connecting the process stages is to assess the flows to be carried. The maximum flow is of most importance but for the conduits which have to carry solids in suspension velocities at flows lower than the maximum will have to be checked so in this case minimum, mean and maximum flows are all needed.

Where all or part of the sewerage system is combined, the maximum flow to be carried from the works inlet to the storm weir will be determined by the storm overflow upstream of the works. Section 4.13 gives the current recommendation for through-flow at first spill. However, the through-flow at maximum spill may be higher and should be estimated.

Below the storm weir, design flows may be obtained from

$$aD + a'E + I$$

in which D, E and I are the mean dry-weather flows of domestic sewage, trade effluent and infiltration water respectively, see p. 10. The multiplier a will be 3 for the maximum flow, 1 for the mean flow and very small, possibly zero, for minimum flow. Flows for $a = 2$ corresponding to the daily maximum will be needed if difficulty is encountered in securing self-cleaning velocities at low flows. Unless the variation of trade effluent during the day is markedly different from that of domestic sewage, a' will be the same as a. The mean and minimum design flows upstream of the storm weir will be the same as those downstream. The maximum flow to the storm tanks is simply the maximum flow into the works less the maximum flow for full treatment.

Design flows from the above formula apply to the whole of the main line from the storm weir to the outfall except for the conduits carrying mixed liquor from aeration tanks to final settling tanks, where $r(D + E)$ must be added in which r is the recycle ratio. When computing maximum flow it may be advisable to base r on the installed capacity of return sludge pumps rather than on the lower normal return rate in case a high rate of recycle is needed occasionally.

For filters operated by recirculation, the recycle rate is often varied so that the ratio to inflow is kept constant. The total flow through the filters and humus tanks must then be expressed as $(1 + r)(aD + a'E + I)$ with r typically equal to one. However, recirculation pump capacity is not always provided to match the highest inflows. Thus if the maximum recirculation capacity is

Fig. 14.2 Aerial view of Ringley Fold Sewage Treatment Works, Bolton (under construction) (Courtesy I. Withnell and G.F. Read of the former Bolton and District Joint Sewerage Board and Amec Croots Babcock Ltd)

say $2(D + E)$, the maximum flow through the filters will be $5(D + E) + I$ with $a = a' = 3$, not $6(D + E) + I$ as it would be if r were left equal to one up to inputs of $3(D + E) + I$.

With alternating double filtration, the whole flow passes through each half of the plant in turn but here again the highest flows may not be fully provided for, single-pass operation or limited recirculation being used whenever the inflow is above a certain level.

When assessing design flows for the individual tanks of a group, mean and minimum flows are obtained by dividing total flow by the number of units, but for maximum design flows one or more units should be considered to be inoperative. With pipe fed units particularly, sharing the maximum flow between fewer units will usually increase head loss considerably. Since this represents an extreme case, freeboard allowances in feed wells and channels can be reasonably reduced to satisfy it.

There are minor recirculations delivered to the main feed channel consisting of surplus secondary sludge, tertiary treatment wash water and supernatant liquor from sludge de-watering. The flows are intermittent but the rates are usually negligible from the point of view of hydraulic design of the main conduits.

Where the sewerage system is rigidly separate the flows will be as described above but with $a = 4$ for maximum flows, I is sometimes considered to be included in $4(D + E)$, provided that this multiplier can truly be considered to give a safe maximum. If any surface water is likely to enter the sewers a higher flow must be taken and consideration will often be given to providing storm tanks, the works then being designed as though it were for a combined system. Since there are unlikely to be storm overflows on the sewers, the maximum flow to the works is not easy to estimate without gaugings.

In tropical climates where maximum rainfall intensities are very high, rain falling directly into process tanks increases maximum flows through the works significantly enough to be added to the sewage flows.

(2) General form of pipes and channels

At all but the smallest works, the tank capacity at each stage of treatment is divided into several units operating in parallel. This avoids the need for standby provision since, whenever a unit is out of service, a proportion of the capacity remains in use, and the nature of the processes is such (Fig. 8.6) that a more than proportionate degree of treatment will still be available.

The channels, pipes and distribution chambers conveying the flow between stages are arranged to divide the flow among the units of each stage and then to collect the flow together again for re-distribution among the units of the next stage, Figs 14.1 and 14.2. If instead the flow once divided passed through the whole sequence of treatment in parallel lines, shut-down of a single unit would render the whole of its line inoperative.

To make the best use of the capacity provided, flow needs to be divided equally between the units at each stage of treatment, especially at times of maximum flow.

Other characteristics which affect sewage works hydraulics are the need to maintain self-cleansing velocities, and the wide range of flows to be dealt with, about 20 to 1 above the storm weir and 10 to 1 below.

The combination of these factors results in a need to avoid the backing up of low flows in feed channels and hence in a need for the top water level (TWL) of each group of tanks to be not higher than the bed of the channel delivering flow to it. Even on a site where there is ample fall, loss of head between stages is of concern because it affects the levels, and hence the excavation volumes, for all subsequent treatment units.

Where pumping between stages is unavoidable it is usually confined to one point, often at the works inlet, but at filter plants, pumping after primary sedimentation can be preferable. There is at least one example of the pumping of mixed liquor to final settlement.

(3) Distribution chambers

Figs 14.3 to 14.5 show some forms of chamber used to distribute flow to circular sedimentation tanks.

In Fig. 14.3, the feed channel terminates in a feed-well from which pipes radiate to four tanks. Each pipe has a penstock to enable flow to be shut off. Equality of flow division depends upon having identical tank weir levels and identical pipe runs serving the tanks. Even then the disposition of pipe inlets in the feed-well may result in some inequality. An octagonal form of well as shown on the lower half of the plan is sometimes thought to be better than a

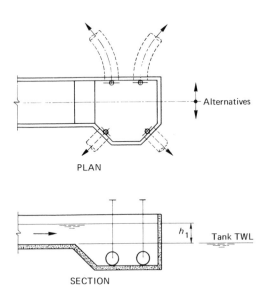

PLAN

SECTION

Fig. 14.3   Channel-fed distribution well

plain rectangular one and eliminates a 45° bend in each pipeline.

Tank TWL is shown as being at the same level as the bed of the feed channel. It could be lower if ample fall was available. Friction loss through the pipes imposes a water level upon the feed-well and hence upon the flow in the channel. A simple design approach, avoiding the necessity to calculate gradually-varied flow, is to calculate a gradient for the channel such that, at maximum flow, normal depth level for the channel is the same as the level imposed at the feed-well. With such a gradient, an M2 (p. 133) profile which need not be calculated, can be relied upon at lower flows, since the water level in the well will then be below normal depth level. Occasionally a design can be improved by using a different bed gradient (perhaps zero) but the implications of gradually-varied depth must then be considered.

On many sites there are advantages in carrying the flow most of the distance to the sedimentation plant by means of a pipe as shown in Fig. 14.4. The symmetry of this form of chamber should assist in equalising the division of flow though identical weir levels and pipe runs to the tanks will still be essential. The figure shows the whole of the head for the tank feed pipes as being provided below the bed level of the channel so that tail water level in the channel depends solely on the head requirement of the main feed pipe. Alternatively, the tank TWL may be at the feed channel bed level, but channel water level will then depend on the sum of the head requirements for both the main feed pipe and the tank feed pipe. Intermediate positions for the TWL are also possible.

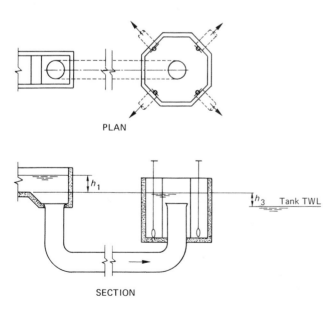

PLAN

SECTION

Fig. 14.4 Pipe-fed distribution chamber

In a chamber of the form shown in Fig. 14.5, equality of flows to the tanks depends solely on the symmetry of the weir slots in the walls of the inner part of the chamber. The tanks can be at different distances from the chamber and even at different levels provided that none of the chamber weirs is drowned. Tank weir levels have therefore to be sufficiently below weir level in the chamber to provide the head requirement for the tank feed pipes. With channel bed level at chamber weir level, tail water level in the feed channel is determined by the sum of the head requirements of the main feed pipe and the weirs.

A group of four filters would be served by a chamber similar to that in Fig. 14.5 but with much shallower peripheral chambers equipped with dosing siphons to ensure that flow passed to a filter only when there was sufficient head to rotate the distributor. For high rate filters, the dosing siphons might not be needed.

Rectangular sedimentation tanks are often fed by pipe systems which

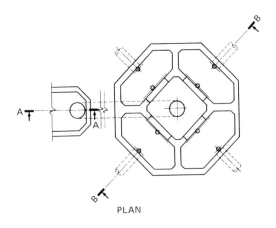

PLAN

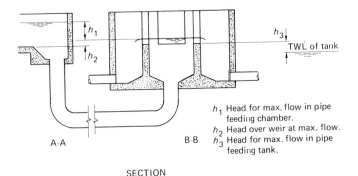

A-A   B-B

h_1 Head for max. flow in pipe feeding chamber.
h_2 Head over weir at max. flow.
h_3 Head for max. flow in pipe feeding tank.

SECTION

Fig. 14.5 Distribution chamber with weirs

bi-furcate successively on a symmetrical plan. A similar principle, using channels is found on aeration tanks but cut-water gates or penstocks, preferably controlled by gauging flumes, become necessary if flexibility of operation is required.

(4) Design velocity for pipes

Where pipes remain full all the time, as in the design shown in Figs 14.3 to 14.5, they are subject to a range of velocity as wide as that of the flow, so for pipes feeding sedimentation tanks, a design velocity must be chosen which will avoid the accumulation of deposits in the pipe. Grease accumulating at the crown is a hazard as well as heavier matter at the invert. A velocity of 1 m/s at maximum flow is usually considered to be the least which can be risked. This implies minimum velocities as low as 0.1 m/s for primary tank feed pipes, which is of course below the 0.3 m/s to which the flow has been subjected in the grit removal tanks. However, the flow will be up to about two-thirds of the maximum for a few hours each day, even in dry weather, so some scouring action is likely.

Design velocities, for maximum flow, higher than 1 m/s should be used where the greater head loss can be tolerated, but it is seldom that a velocity in excess of about 1.6 m/s can be used, head loss increasing approximately as the square of velocity.

(5) Design velocity and depth/breadth ratio for channels

Channels carrying liquor with particles in suspension still need self-cleansing velocities but variation of flow poses a less serious problem than with pipes since depth of flow changes with changes in discharge. Velocity in a freely discharging rectangular channel may still be two-thirds of its value at maximum flow when the discharge has been reduced to one-fifth.

The choice of design velocity and depth/breadth ratio for a channel influence its maximum depth of flow which in turn has a major effect on the difference in TWL of the tanks for successive treatment stages. However, high velocities, though reducing depth, require steeper bed gradients if designed for normal depth flow or will have steeper surface slopes if the bed is level. For small works the effect of slope is as significant as that of depth and, where head is limited, design velocity may need to be kept down to about 1.0 m/s.

On larger works, higher design velocities are often beneficial, enabling channels to be of smaller cross-section as well as reducing the successive steps in top water levels. Nevertheless, if too high a velocity is adopted, afflux at converging junctions (p. 141), especially T-junctions, increases upstream channel depths markedly. As with pipes, velocities more than about 1.5 m/s seem rarely feasible.

Keeping the depth/breadth ratio low presents another way of limiting depth. A ratio of 0.5 is perhaps an economic choice in many cases. The need for freeboard above the maximum water level should not be forgotten when arriving at overall dimensions.

At works where the flow arrives in ordinary sewers, the size of sewer influences the depth of channels between the works inlet and the primary sedimentation tanks. This is discussed further in relation to the case for which calculations are given in the next section.

14.5 Illustrative hydraulic calculations

In order to show how some of the principles described in Chapters 6 and 7 can be used in designing pipe and channel systems some illustrative calculations are given below.

They relate to a works serving a population of 50 000 with a mean dry weather flow rate of 180 litre per head per day, giving a mean domestic flow, D, of 9 Ml/d. Mean industrial effluent flow, E, and infiltration, I, are taken as 2.0 and 2.5 Ml/d respectively, giving

$$DWF = D + E + I = 13.5 \text{ Ml/d}$$

and the maximum flow for full treatment as:

$$\text{'3 DWF'} = 3(D+E) + I = 35.5 \text{ Ml/d} = 0.411 \text{ m}^3/\text{s}$$

Flow gaugings would form the basis of the data relating to all three components of flow wherever possible.

The first two columns of Table 14.1 contain calculations for the head losses in 600 and 700 mm nominal diameter pipes to be considered as alternatives for the pipeline which is to carry the flow from the main feed channel to a distribution chamber located at the centre of a group of four primary sedimentation tanks. The diameters were chosen as the ones giving velocity at maximum flow (see the fourth line) of between 1.0 and 1.5 m/s. Actual diameters are a few percent larger than nominal diameters so the head losses will be over-estimated.

The fifth line of the table is to confirm that the rough-turbulent formula is valid for the calculation of the friction factor, f, which is given in the seventh line.

The nett length of pipe is given in the eighth line and the total equivalent length is given in the eleventh line, 85 diameters having been added to allow for the entrance loss (25) and two right-angled bends (30 each).

In the formula for head loss on the bottom line, the first term in the bracket allows for the kinetic energy remaining in the flow as it leaves the pipe, hardly any of which is recoverable as pressure energy. Alternatively, $1/(4f)$ diameters may be added to the length.

Calculations for alternative diameters of feed pipes for the individual tanks are given in the third and fourth columns of Table 14.1 and follow the procedure just described. The 65 diameters of pipe added to the nett length

Table 14.1

		Main feed pipe		Tank feed pipe	
Q	m³/s	0.411	0.411	0.103	0.103
d	mm	600	700	300	350
k_s	mm	1.5	1.5	1.5	1.5
V	m/s	1.45	1.07	1.45	1.07
$Vk_s > 9 \times 10^{-4}$	m²/s	2×10^{-3}	1.6×10^{-3}	2×10^{-3}	1.6×10^{-3}
d/k_s	mm/mm	400	467	200	233
f from $\dfrac{1}{\sqrt{f}} = 4 \log (3.7d/k_s)$		0.00622	0.00596	0.00759	0.00725
Length	m	25	25	20	20
Entrance, bends etc. Nd		85d	85d	65d	65d
	m	51	59.5	19.5	22.75
Total equivalent length	m	76	84.5	39.5	42.75
$h = \dfrac{V^2}{2g}\left(1 + \dfrac{4fL}{d}\right)$	m	0.460	0.225	0.539	0.263

allow for loss at the entrance (25), two $22\frac{1}{2}°$ bends (5 each) and a 90°
bend (30).

If a distribution chamber like that in Fig. 14.5 is to be used account will
have to be taken of the head over the weir sills. Taking a fairly low value of
the coefficient of 1.7 m$^{1/2}$/s, and a weir length of 1.0 m, the head, h_w is given
by:

$$0.103 = 1.7 \times 1.0 \times h_w^{3/2}$$

From this h_w will be 0.154 m.

Choices now have to be made of the type of distribution chamber and the
diameters of the pipes but these decisions will depend on the depth of flow
available in the main feed channel as affected by the diameter of the incoming
sewer and by the sill height and operating head for the storm weir.
Figure 14.6 shows changes in depth between the sewer and the sedimentation tank feed-well. The total change is made up of: head loss at the screens
before cleaning, at least 26% of upstream depth at each gauging flume unless
it has a long divergence (see Section 7.7), and the head over the storm weir at
maximum flow. The depths shown on Fig. 14.6 are consistent with a pipe of
600 mm feeding a chamber of the form shown in Fig. 14.5 with weirs 1.0 m
long. The weir sills are at the same level as the bed of the feed channel so the
depth imposed upon the downstream end of the channel by the pipe and the
distribution chamber consists of the sum of the head over the weir and the
loss through the pipe. Top water level of the sedimentation tanks will be
below chamber weir sill level to the extent of the head loss in the pipes feeding
the individual tanks plus a small freeboard allowance.

The choice of diameter for the pipes feeding the tanks will depend on the
availability of fall through the whole works and may be deferred until other
parts of the works have been investigated. The larger diameter would be

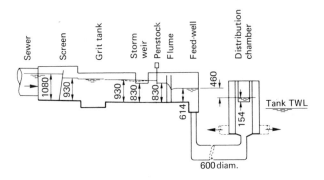

Fig. 14.6   Diagrammatic profile from sewer outfall to primary tanks. The channel bed is shown level from the storm weir to the gauging flume. Elsewhere the slopes could be for normal depth at maximum flow, locally steeper at transitions. If a measuring flume were required upstream of the storm weir, the depths would have to be re-considered. Total flow could be measured alternatively by providing a flume in the storm tank feed channel

chosen if fall were limited. With ample fall available, consideration can be given to designing the pipes and channels to carry the full maximum flow when only three tanks are operating. This will increase the flow per tank by a factor of $\frac{4}{3}$ and hence the head loss in a given diameter of pipe by about $(\frac{4}{3})^2$. If, as in the calculation illustrated, an inoperative tank is not allowed for the storm tanks might have to be used for temporary storage whenever the flow exceeded about $\frac{3}{4}$ of the maximum with one tank inoperative. Any lowering of the sedimentation tanks judged to be feasible would increase the fraction of maximum flow which could be treated fully with a tank shut down.

The system of channels for the four tanks is shown in Fig. 14.7 and is designed for a velocity of about 1.0 m/s and a depth/breadth ratio of 0.5, common to all channels. These are not necessarily the best values to choose and the reader may wish to try others in order to see their effects on dimensions and slopes.

Calculations for the several channels are shown in Table 14.2, the columns taking the channels in order moving upstream. The procedure for the peripheral channel, which has spatially varied flow, differs from that for the other channels but many of the steps of calculation are the same so it is included in the table.

With four tanks, the flows in successive channels, moving upstream, are halved as shown. The lengths in the second line come from a layout plan and are consistent with tanks of 20 m diameter. A roughness value of 1.5 mm has been used, though advocates might be found for 3.0 mm or 0.6 mm and the reader may wish to try the effect of these alternatives. The breadths and depths have been calculated to meet the objectives of $V = 1$ m/s and $y/b = 0.5$, with some rounding of dimensions.

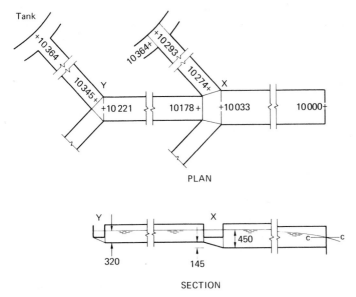

PLAN

SECTION

Fig. 14.7 Sedimentation tank collector channels

Table 14.2

		X to outfall	Y to X	Tank to Y	Peripheral
Q	m³/s	0.411	0.205	0.103	0.0514
L	m	25	22	6	32
k_s	mm	1.5	1.5	1.5	1.5
b	m	0.90	0.64	0.45	0.32
y	m	0.45	0.32	0.225	0.313
V	m/s	1.015	1.001	1.017	0.513
$Vk_s > 9 \times 10^{-4}$	m³/s	1.5×10^{-3}	1.5×10^{-3}	1.5×10^{-3}	8×10^{-4}
$\mathscr{F}^2 = V^2/(gy)$		0.233	0.319	0.469	0.0857
$m = by/(b + 2y)$	m	0.225	0.160	0.1125	0.106
$d/k_s = 4\,m/k_s$	m/m	600	427	300	282
f from $\dfrac{1}{\sqrt{f}} = 4\log(3.7d/k_s)$		0.00558	0.00610	0.00674	0.00686
$s = fV^2/(2gm_n)$		1/769	1/514	1/316	
$\Delta z = sL$	mm	32.5	43	19	

The method adopted is to find by calculation the gradients which will make the chosen depths into the normal depths, y_n, thus avoiding a need to calculate water surface profiles. If there is free discharge at the outfall of the main channel, depth at that point will be close to critical depth, y_c, and there will be an M2 profile upstream, asymptotic to normal depth. It is assumed that the channel is long enough for depth to be negligibly different from normal when the junction X is reached. Incidentally, depth imposed by downstream conditions at the outfall end can be as high as normal depth without making the depth at X more than the normal.

The purpose of the seventh line of the table is to check that flow is in the rough-turbulent zone. The next line gives $\mathscr{F}^2$ which will be needed when the junctions are considered. The hydraulic mean depth, m, is needed (line 9) and the roughness ratio (line 10) which is $4m/k_s$ for a non-circular section.

The rough-turbulent zone formula is used for f. The bed gradient, s, to make the chosen depth normal is then calculated from the Darcy–Weisbach formula.

Finally the fall, Δz, along the channel is found.

The step heights at the junctions remain to be determined and for these the formulae in Section 7.5 have been used. Changes in bed level at the junctions will be in the form of short steeply sloping transitions but the basic equations for these are the same as for abrupt vertical steps.

At junction X, using equation (7.17),

$$y_1 + h = 0.45(1 + 2 \times 0.233(1 - 0.5 - 0.5\cos 45°))^{1/2} = 0.465 \text{ m}$$

and since we have already decided to make $y_1 = 0.32$, the step height, h, becomes 0.145 m. The side branch step becomes $0.465 - 0.225 = 0.240$ m.

At junction Y, using equation (7.18),

$$y_1 + h = 0.32(1 + 2 \times 0.319(1 - \cos 45°))^{1/2} = 0.349 \text{ m}$$

giving a step height of $0.349 - 0.225 = 0.124$ m.

At the head of the branch channel, using equation (7.19),

$$y_1 + h = 0.225(1 + 2 \times 0.469)^{1/2} = 0.313 \text{ m}$$

The branch channels above Y join the peripheral channels of the sedimentation tanks without a step. The depth 0.313 m is therefore taken as the depth at the downstream end of the peripheral channels and is shown as such in the fourth column (line 5) of the table. Depth will vary over this cross-section and the average depth may be a little less than 0.313 m which is the estimate of the maximum depth. Using rounded rather than right-angled corners at this junction would probably be beneficial in reducing the upstream depth.

The peripheral channels of the downstream pair of tanks do step down to the branch channels but these steps are a consequence of levels already fixed by the step at junction X and the slope of the branch channel, and does not require a flow calculation.

The breadth of the peripheral channel, 0.32 m, has been chosen so that the channel will not be awkward to clean. A narrower channel would have been cheaper structurally and still satisfactory hydraulically.

The fourth column of the table provides the means of calculating Li's $\alpha = fL/(2m_2)$ and the a and b of equation (7.14), p. 140. The value of Vk_s showed that flow was just in the transitional zone so f was increased to 0.007

in calculating α. The values of α, a and b are 1.057, 2.792 and 0.453 respectively giving for zero slope a depth of 0.345 m in the channel at the furthest point, diametrically opposite to the outlet.

With a slope of, say, 1/200, the parameter G becomes 0.511 and, assuming we can legitimately combine equations (7.14) and (7.15), the depth becomes 0.196 m. Adding the slope drop of 3 mm × 1/200 = 0.160 m gives 0.356 m to compare with the 0.345 m without slope. Peripheral channels are often given a slope, usually by screeding the bed after construction, but there seems no strong need for this to be done.

It is prudent to allow 50 or 100 mm of freeboard above the calculated water levels when determining the weir level of the tanks. Depth of flow over plain outlet weirs is negligible, but Vee-notch weirs, commonly used on final settling tanks, have heads at maximum flow of 20 to 40 mm and this could be allowed for when considering levels in upstream distribution chambers and feed channels.

Reference

White, J.B., Aspects of the hydraulic design of sewage treatment works, *The Public Health Engineer*, **10**, **3**, 164–170, July 1982 and **11**, **1**, 58, Jan. 1983.

Index

(Italicised page numbers refer to bibliographical references)